The ACS
Style Guide

The *ACS* Style Guide

A Manual for Authors and Editors

Janet S. Dodd, Editor

Marianne C. Brogan, Advisory Editor

AMERICAN CHEMICAL SOCIETY
WASHINGTON, DC 1986

Library of Congress Cataloging in Publication Data

The ACS style guide.
  Bibliography: p.
  Includes index.

  1. Chemical literature—Authorship—Handbooks,
manuals, etc. 2. English language—Style—Handbooks,
manuals, etc.

  I. Dodd, Janet S., date. II. Brogan, Marianne C.
III. American Chemical Society.

QD8.5.A25   1986   808'.06654   85-21472
ISBN 0-8412-0917-0
ISBN 0-8412-0943-X (pbk.)

Contributors

Janet S. Dodd
Managing Editor
American Chemical Society Books Department
1155 16th Street, NW
Washington, DC 20036

Marianne C. Brogan
Associate Head
American Chemical Society Journals Department
2540 Olentangy River Road
Columbus, OH 43210

Barbara Friedman Polansky
Copyright Administrator
American Chemical Society
1155 16th Street, NW
Washington, DC 20036

Larry Venable
Fred Pryor Seminars
2000 Johnson Drive
Shawnee Mission, KS 66201

Arthur A. Antony
Chevron Research Company
576 Standard Avenue
Richmond, CA 94802

Contents

Figures

Other Illustrations

Lists

Tables

Foreword

More than ten years elapsed between the publication of our first *Handbook for Authors* in 1967 and our second edition in 1978. A further eight years has passed, and now we present this book, which is not intended to be just an updated *Handbook*, but much more—a true "style guide".

Our previous *Handbooks* were devoted almost entirely to instruction of contributors to ACS publications in "the way we do it at ACS". *The ACS Style Guide* does not forget that ACS publications still have their own special requirements and peculiarities, and does include that information. However, the approach taken here is a more general one, stressing those principles and practices that are desirable throughout the scientific literature.

This book also expands the scope of previous *Handbooks* by covering the newly emerging issue of machine-readable manuscripts (Chapter 5); an overview of the chemical literature, of which ACS publications are only a part (Chapter 6); and oral presentation of science (Chapter 7). Such information is both complementary to and different from that required to prepare a written paper.

In short, what you have in your hand is the result of a project that was both larger and more complex than that underlying our previous *Handbooks*. Editor Janet Dodd, Managing Editor in the ACS Books Department, bore the considerable responsibility for compiling this guide. She received valuable aid and advice from Marianne Brogan, Associate Head in the ACS Journals Department, who had shepherded the 1978 version to completion. The Editors are to be commended on their achievement. We at ACS look forward to receiving your comments as readers and users.

D. H. MICHAEL BOWEN
Director, Books & Journals Division

Preface

Most published material, scientific and literary, is the result of dozens, even hundreds, of editorial decisions. One major purpose of this book is to present guidelines for such decisions. The book is addressed to both authors and editors because, essentially, authors start the process and editors finish it. Thus, the chapters and appendixes cover most aspects of the publishing process.

Oscar Wilde said "The difference between journalism and literature is that journalism is unreadable and literature is unread." Of course, scientific writing is quite different from journalism and literary works, but there are similarities, the most important of which are that editors and authors need each other, they need to cooperate, and they need clear guidelines. At ACS, we would like to achieve good author–editor relationships so that we will publish readable and well-read scientific literature. We hope that this book will help us to accomplish that goal.

Acknowledgments

First, I am most grateful to M. Joan Comstock, Head of the ACS Books Department, for giving me the opportunity to work on such a challenging project as this book. Also, it was my pleasure to work with Marianne Brogan, and I am grateful for all I have learned from her. Finally, but no less important, I am grateful for the suggestions of my colleagues who took time to review the chapters or provide useful information: Paula J. Barry, Paula M. Bérard, Stuart A. Borman, M. Joan Comstock, Leroy Corcoran, R. Michael Dodd, Janice L. Fleming, Rani Anne George, Cyrelle K. Gerson, Robin Giroux, Alan Kahan, Hilary M. Kanter, Frances Reed Lyle, Meg Marshall, Karen McCeney, Dolores A. Miner, Gail M. Mortenson, Barbara Friedman Polansky, Martha Polkey, Anne T. Riesberg, Susan Robinson, Elizabeth Shank, Deborah H. Steiner, Marcia Vogel, Louise Voress, and Mary Warner. The chapters were also reviewed by selected members of the Society Committee on Publications: Barbara Wood, Royce Murray, John Nelson, Louis Quin, and Peter Rabideau.

JANET S. DODD

1
Chapter

The Scientific Paper

This chapter is intended to be a general guide to writing a scientific paper. Certain style points and other requirements differ from journal to journal or publisher to publisher. For ACS publications, it is important to consult the Guide, Notes, or Instructions for Authors that appear in each journal's first issue of the year, or the Requirements for Manuscripts for books.

Getting Started

M ost guides to technical writing begin by stating that no set of "writing rules" exists, to be followed like a cookbook recipe or an experimental procedure. However, some guidelines can be helpful, and the first is that the writer start by answering some questions.

The most important questions to answer are the most obvious:

- What is the function or purpose of this paper? Are you describing original and significant research results? Are you reviewing the literature? Are you providing an overview of the topic? Something else?

- How is your work different from other reports on the same subject? (Unless you are writing a review, be sure that your paper will make an original contribution. Some publishers, including ACS, do not permit publication of previously published material.)

- Where is the best place for this paper to be published—in a journal or as part of a book? If in a journal, which journal is most appropriate? (Appendix I, "ACS Publications", describes ACS journals and books.)

- Who is the audience? What will you need to tell them that they don't already know?

Answering these questions will clarify your goals and thus will make it easier for you to write the paper with the proper amount of detail and will also make it easier for editors to determine the paper's suitability for their publications. Writing is like so many other things: if you clarify your overall goal, the details fall into place.

Once you know the function of your paper and who its audience is, review your material for completeness or excess. Then, organize your material into the most standard format: introduction, experimental details or theoretical basis, results, discussion, and conclusions. This format has become standard because it fits most reports of original research, it is basically logical, and it is easy to use. The reason it fits most reports of original research is that it parallels the scientific method of deductive reasoning: define the problem, create a hypothesis, devise an experiment to test the hypothesis, conduct the experiment, and draw conclusions. Furthermore, this format enables the reader to understand quickly what is being presented and to find specific information easily. This ability is crucial now more than ever because most scientists, if not everyone, must read much more material than their time seems to allow.

Even if your results are more suited to one of the shorter types of presentation, the logic of the standard format applies, although you might omit the standard headings themselves or do without one or more entire sections. As you write, you can modify, delete, or add sections and subsections as appropriate.

Check the specific requirements of the publication you have targeted and follow them. Finally, start writing.

Writing Style

Short declarative sentences are easiest to write and easiest to read, and they are usually clear. However, too many short sentences in a row can sound abrupt or monotonous. It is easier to start with simple declarative sentences and then combine some of them than to start with long rambling sentences and then try to shorten them.

By all means, you should write in your own personal style, but keep in mind that scientific writing is not literary writing. Scientific writing serves a completely different purpose from literary writing, and it must therefore be much more precise. Some specific hints follow.

- Stick to the original meanings of words; don't use a word to express a thought if such usage is the fourth or fifth definition in the dictionary or if such usage is primarily literary. Examples are using "since" when you mean "because", and "while" when you mean "although". Many words are clear enough when you are speaking because you can amplify your meaning with gestures, expressions, and vocal inflections—but when written, they're clear only to you.

- Avoid slang and jargon.

- Use strong verbs; they are essential to clear, concise writing.

- Use the active voice whenever possible. It is usually less wordy and unambiguous.

  Poor
  The fact that such processes are under strict stereoelectronic control is demonstrated by our work in this area.

  Better
  Our work in this area demonstrates that such processes are under strict stereoelectronic control.

- Be brief. Wordiness usually adds nothing but confusion, and the resulting paper is very expensive to typeset and to print.

- First person is perfectly acceptable where it helps keep your meaning clear:

  Jones reported xyz, but we found...
  Our recent work demonstrated...
  For these reasons, we began a study of...

  However, phrases like "we believe", "we feel", "we concluded", and "we can see" are unnecessary, as are personal opinions.

- Try not to shift verb tenses within the same paragraph and section. However, the tense should change from section to section. Present and past tenses are correct in the introduction: "Absolute rate constants for a wide variety of reactions are available. Jones reviewed the literature and gathered much of this information". Simple past tense is correct for describing your procedures: "The solutions were heated to boiling", "the spectra were recorded". Then use present tense to discuss your results and conclusions.

- If you have already presented your results at a symposium or other meeting and are now writing the paper for publication in a book or journal, delete all references to the meeting or symposium such as "Good afternoon, ladies and gentlemen", "This morning we heard", "in this symposium", "at this meeting", "I am pleased to be here." Such phrases would be appropriate only if you were asked to provide an exact transcript of a speech.

  One final point is that you and those around you probably have been discussing the project for months, so the words seem familiar, common, and clear to you and your colleagues. That's where editors can help. Chapter 2, "Grammar, Style, and Usage", is primarily directed to editors.

Components of Your Paper

The standard format is appropriate for reports of original research but not necessarily for literature reviews or theoretical papers.

Title

The best time to determine the title is after you have written the text, because the title should reflect the paper's content and emphasis accurately and clearly. Choose terms that are as specific as the text permits: "a vanadium-iron alloy" is better than "a magnetic alloy". The title must be brief and grammatically correct but accurate and complete enough to stand alone. A two- or three-word title may be too vague, but a fourteen- or fifteen-word title is probably unnecessarily long. Avoid phrases like "on the", "a study of", "research on", "report on", "regarding", and "use of". In most cases, omit "the" at the beginning of the title. Avoid nonquantitative, meaningless words like "rapid" and "new". In books but not in journals, if you cannot come up with a title that is short enough, consider breaking it into title and subtitle.

The title should attract the potential audience and aid retrieval and indexing services, so it is important to include several keywords.

Series titles are of little value. Some publications do not permit them at all. Consecutive papers in a series, published simultaneously, are generally sufficiently closely related so that a series title may be relevant, but paper 42 probably bears so limited a relationship to paper 1 that it does not warrant parallelism. In addition, unnecessary duplication of the title may cause the editor or reviewer to be prejudiced against the paper and consider it only one more publication on a general topic that has already been discussed at length.

Spell out and define each new term, and avoid jargon, symbols, formulas, and abbreviations. Whenever possible, use words rather than expressions containing superscripts, subscripts, or other special notations.

Do not cite company names or specific trademarks of instruments or materials in titles.

Byline and Affiliation

Include in the byline all those, and only those, who have made substantial contributions to the work, even if the paper has actually been written by only one person.

Use first names, initials, and surnames (e.g., John R. Smith) or first initials, second names, and surnames (e.g., J. Robert Smith). Do not use only initials with surnames (e.g., J. R. Smith) because this style causes indexing and retrieval difficulties and interferes with unique identification of an author.

Whatever byline you use, be consistent. Papers by J. Smith, J. R. Smith, John R. Smith, Jr., and J. R. Smith, Jr., will not be indexed in the same place; the bibliographic citations may be scattered at four different locations, and ascribing the work to a single author will therefore be difficult.

Do not include professional or official titles or academic degrees.

The first or last author is generally considered the senior author. Unless there is only one author, indicate the author to whom correspondence should be addressed with an asterisk or superscript (check the specific publication's style). The affiliation should be the institution where the work was conducted. If the present address of an author differs from that at which the work was done, give the current address in a footnote.

Abstract

Many publications require that an informative abstract accompany every paper. For a research paper, the abstract should summarize the principal findings; for a review paper, the abstract should describe the topic, the scope, the sources reviewed, and the conclusions. You should write the abstract last to be sure that it reflects accurately the content of the paper.

The purposes of the abstract are to allow the reader to determine the nature and scope of the information given in the paper and to allow editors to pinpoint key features for use in indexing and eventual retrieval. The abstract must contain sufficient information to allow a reader to decide whether to read the whole paper. The abstract should provide adequate data for generating index entries. Although an abstract should not be a substitute for the article itself, it must be complete enough to appear separately in abstract publications. Often, authors' abstracts are used directly in *Chemical Abstracts*.

In the abstract you should briefly state the problem or the purpose of the research when that information is not adequately contained in the title, indicate the theoretical or experimental plan used, accurately summarize the principal findings, and point out major conclusions. Include chemical safety information when applicable. Do not add to, evaluate, or comment on conclusions in the text.

The abstract should be concise and self-contained. The optimum length could be two sentences; it could be many more, depending on the subject matter and the length of the paper. Use meaningful nomenclature; that is, use standard systematic nomenclature where specificity and complexity require, or use trivial nomenclature where it will adequately and unambiguously define a well-established compound.

Do not cite references, tables, figures, or sections of the paper. You may refer to equations or structures presented in the body of the paper if they are on a single line and can readily be incorporated when the abstract is used in the secondary literature (e.g., *Chemical Abstracts*). Do not include equations and structures that take up more than one line.

Use abbreviations and acronyms sparingly and only when necessary to prevent awkward construction or needless repetition. Define abbreviations at first use in the abstract (and again at first use in the text).

Introduction

A good introduction is a clear statement of the problem or project and why you are studying it. This information should be contained in the first few sentences. On the basis of your answer to the question of what the readers already know, give a concise background of the problem and the significance, scope, and limits of your work. Outline what has been done before by citing truly pertinent literature, but do not include a general survey of semirelevant literature. State how your work differs from work previously published or state how it is related. Demonstrate the continuity from the previous work to yours. Your introduction should be at least one or two paragraphs long. Ordinarily, the heading "Introduction" is not used because it is superfluous; opening paragraphs are usually introductory in nature.

Experimental Details or Theoretical Basis

In research reports, this section can also be called "Experimental Methods", "Experimental Section", or "Materials and Methods". Check the specific publication. For experimental work, give enough detail about your materials and methods so that other experienced workers could repeat your work and obtain comparable results. When using a standard method, reference the appropriate literature and give only the details needed.

Identify the materials you used, and give information on the degree and criteria for purity, but do not reference standard laboratory reagents. Give the chemical names of all compounds and the chemical formulas of compounds that are new or uncommon.

Describe your apparatus only if it is not standard and not commercially available. Giving a company name and model number in parentheses is adequate and nondistracting.

Avoid using trademarks and brand names of equipment and reagents. Use generic names; include the trademark in parentheses after the generic name only if the product you used is somehow different from others. Furthermore, trade names often are recognized and available as such only in the country of origin.

Describe the procedures you used, unless they are established and standard.

Note and emphasize any hazards, such as explosive tendencies and toxicity, in a separate paragraph introduced by the word "Caution". Include precautionary handling procedures and any other safety considerations. Some ACS journals and books will also indicate the hazard as a footnote on their contents pages.

In theoretical reports, this section would be called "Theoretical Basis" instead of "Experimental Details" and should include sufficient mathematical detail to enable derivations to be reproduced and numerical results to be checked. Include all background data, equations, and formulas necessary to the arguments. Place lengthy derivations in an appendix or in supplementary material.

Results

Summarize the data collected and the statistical treatment of them. Include only relevant data, but give sufficient detail to justify your conclusions. Use equations, figures, and tables only where necessary for clarity and conciseness. (See Chapter 3, "Illustrations and Tables".)

Discussion

When discussing your results, be objective. Point out the features and limitations of the work, and interpret and compare your results. Relate your results to your original purpose in undertaking the project: Have you resolved the problem? What exactly have you contributed? Briefly state the logical implications of your results. Suggest further study or applications if warranted.

You can present your results and discussion as two separate sections, or you can combine them into one section if it is more logical to do so.

Conclusions

If you have already presented your conclusions in your "Discussion" section, you do not need a separate section for them. If you do have a separate "Conclusions" section, do not repeat discussion points or include irrelevant material. Your conclusions should be based on the evidence presented.

Summary

A summary is unnecessary in short papers (less than 20 typed double-spaced manuscript pages) and many longer papers. In longer papers a summary of the main points can be helpful, but only the main points should be included. If the summary is too long, its purpose is defeated.

Acknowledgments

Check the style of the publication for which you are writing, but generally the last paragraph of the paper is the place to acknowledge people, places, and financing. As simply as possible, thank those persons, other than coauthors, who have added substantially to the work or who have aided materially by providing equipment or supplies. Do not include their titles. You may state grant numbers and sponsors here. Some journals put financial aid and meeting references together, but not in the acknowledgments section. Some journals permit financial aid to be mentioned in acknowledgments, but not meeting references.

References

In many journals and books, references are placed at the end of the article or chapter; in others, they are treated as footnotes. In ACS books and most journals, the style and content of references are standard regardless of where they are located. Submit your list of references typed double spaced at the end of the paper. Follow the reference style presented in Chapter 2.

Special Sections

The preceding discussion on format applies to most manuscripts, but it is not intended as a set of rigid rules and headings. As long as your paper is well organized, scientifically sound, and appropriate to the publication for which you are preparing it, other sections and subsections may be included.

Sections that are often, but not always, useful or helpful are List of Abbreviations, List of Mathematical Notation, List of Symbols, Glossary, Supplementary Material paragraph, and Appendix. These sections are useful with longer papers (more than 40 typed double-spaced pages). An appendix should contain material that is supplementary to the content of the text, provides important background information, but would disrupt the logic of the presentation.

Types of Presentations

The following are general descriptions; Appendix I discusses each type of presentation with specific reference to ACS publications.

Articles

Articles are definitive accounts of significant, original studies. They should present important new data or provide a fresh approach to an established subject.

The organization and length of these contributions should be determined by the amount of new information and data to be presented and by space restrictions within the publication. The standard format is suitable for most papers in this category.

Notes

Notes are concise accounts of original research of a limited scope. They may also be preliminary reports of special significance. The material reported must be definitive and may not be republished elsewhere later. Appropriate subjects for notes are improved procedures of wide applicability or interest, accounts of novel observations or of compounds of special interest, and development of a new technique. Notes are subjected to the same editorial appraisal as full-length articles.

Communications

Communications, called "Letters" or "Correspondence" in some publications, can be preliminary reports of special significance and urgency that are given expedited publication. They will be accepted if, in the opinion of the editors, their rapid publication will be a service to the scientific community. They may also be comments on the work of others, in which case the original authors' rebuttal may be published at the same time. They are subject to length limitations, and they must contain specific results to support their conclusions, but may contain no polemics or nonessential experimental details.

Communications are submitted to review, and they are not accepted if, in the opinion of the editor, the principal content has been published elsewhere. The same rigorous standards of acceptance that apply to full-length papers also apply to communications. In many cases, authors are expected to publish complete details (not necessarily in the same journal) after their communications have been published. Acceptance of a communication, however, does not guarantee acceptance of the detailed manuscript.

Reviews

Reviews integrate, correlate, and evaluate results from published literature on a particular subject. They seldom report new experimental findings. Effective review articles have a well-defined theme, are usually critical, and should present novel theoretical interpretations. Ordinarily they do not give experimental details, but in special cases (as when a technique is of central interest) experimental procedures may be included. An important function of reviews is to serve as a guide to the original literature; for this reason accuracy and completeness of references cited are essential. Reviews should critically analyze the literature.

Book Chapters

In multiauthored books, chapters may be, like journal articles, accounts of original research or literature reviews, but they may also be topical overviews. They may be developed and expanded from presentations given at symposia, or they may be written especially for the book in which they will be published. Multiauthored books preferably contain at least one chapter that reviews the subject thoroughly and also provides an overview; these unify the chapters into a coherent treatment of the subject. In a longer book that is divided into sections, each section may need a short overview chapter.

In books entirely written by one or two authors, each chapter treats one subdivision of the broader topic, and each is a review and overview.

2
Chapter

Grammar, Style, and Usage

Grammar, style, and usage are the copy editor's specialties. Copy editors are detail-oriented; authors are oriented more toward their scientific results. Copy editors are trained to standardize the numerous details of style, to find grammatical errors, and to smooth over awkward or unclear usage. Although ACS copy editors perform these functions on all manuscripts, this chapter is not intended only for copy editors; it is also addressed to authors. The more the manuscript conforms to ACS style, the faster it can be processed.

Many authors ask why we have a style for seemingly trivial elements like capitalization, hyphenation, abbreviations, and so on. Why can't each author do it his or her own way? A consistent style provides unity and coherence to the journal or book and makes communication clear and unequivocal; thus it saves readers time and effort by not allowing a variety of styles for the same thing to distract them from the content. If readers must pause, even for a moment, to think about matters of style, it will take a lot longer to read the article. A consistent style also provides transportability: a paper submitted to one journal is more easily resubmitted to another, more specialized or appropriate, publication. A consistent style also ensures similarity of treatment of papers tailored for different publications.

Style is like language: it evolves, adapting to change by reacting appropriately to new challenges, by discarding that which has outlived its usefulness, by questioning current practice and reaffirming or challenging its value.

This chapter will present grammatical errors that ACS copy editors see frequently. It will not attempt to cover all the rules of grammar; many excellent grammar texts are available for that purpose. Stylistic or editorial conventions, mainly but not solely for technical material, are also presented in this chapter.

The style adopted by ACS is for the most part taken from established authoritative sources, such as *The Chicago Manual of Style*, *Words into Type*, and the *Government Printing Office Style Manual*.

0917-0/86/0011$21.20/0   © 1986 American Chemical Society

Grammar

Subject–Verb Agreements

Everyone knows that a subject and its verb must agree in number. However, subject–verb disagreements are quite common errors. Often, one or more prepositional phrases between the subject and the verb cause this error.

Incorrect
Application of this technique to studies on the phytoplankton biomass and its environments are described. (The subject is "application", which is singular.)

Correct
Application of this technique to studies on the phytoplankton biomass and its environments is described.

Sometimes, two singular subjects joined by "and" cause this error.

Incorrect
Growth and isolation of M13 virus was described.

Correct
Growth and isolation of M13 virus were described.

Exception: A subject that is plural in form but singular in effect takes a singular verb.

Examples
research and development   name and address

Research and development is attracting a growing number of young scientists.

The name and address of each contributor is given on the title page.

However, when two or more subjects are joined by "or", the verb takes the number of the closest subject.

Correct
Application or uses were noted.

Also correct
Uses or application was noted.

Sometimes, determining the number of the subject is difficult. Be aware of collective nouns that take a singular verb. Collective nouns take a singular verb when the group as a whole is meant; they take a plural verb when individuals of the group are meant.

Examples

contents	dozen	none	range
couple	group	number	series
data	majority	pair	variety

Incorrect
The series are arranged in order of decreasing size.

Correct
The series is arranged in order of decreasing size. (Refers to the series as a unit.)

Incorrect
A series of compounds was tested.

Correct
A series of compounds were tested. (Refers to each compound.)

Incorrect
Experimental data that we obtained is compared with previously reported results.

Correct
Experimental data that we obtained are compared with previously reported results. (Refers to the data as individual results.)

Incorrect
After the data are distributed, we can meet to discuss them.

Correct
After the data is distributed, we can meet to discuss it. (Refers to the whole collection of data as one unit.)

Incorrect
None of the samples was soluble.

Correct
None of the samples were soluble. (Refers to individuals.)

Incorrect
This group of workers are well aware of their responsibilities.

Correct
This group of workers is well aware of its responsibilities. (Refers to the group as a unit.)

Incorrect
This group of workers is willing to sign their names.

Correct
This group of workers are willing to sign their names. (Refers to the individuals.)

Units of measurement are treated as collective nouns and therefore take a singular subject.

> *Incorrect*
> The mixture was stirred, and 5 mL of diluent were added.
>
> *Correct*
> The mixture was stirred, and 5 mL of diluent was added.
>
> *Incorrect*
> Five grams of NaCl were added to the solution.
>
> *Correct*
> Five grams of NaCl was added to the solution.
>
> *Incorrect*
> Three weeks are needed to complete the experiment.
>
> *Correct*
> Three weeks is needed to complete the experiment.

Compound subjects containing the words "each", "every", and "everybody" may take singular verbs.

> *Examples*
> Each flask and each holder was sterilized before use.
>
> Every rat injected and every rat dosed orally was included.
>
> Everybody in the group and every visitor is assigned a different journal each month.

Both components of the compound subject must contain the words in question. Otherwise, the verb must be plural.

> *Example*
> Each student and all of the professors were invited.

Restrictive and Nonrestrictive Clauses

When a clause in a compound sentence is restrictive, the clause is necessary to the sense of the sentence; the sentence would become pointless without the clause. Restrictive clauses are best introduced by "that", not "which".

> *Incorrect*
> It was necessary to find a blocking group which would react with the amino group but not with the hydroxyl.
>
> *Correct*
> It was necessary to find a blocking group that would react with the amino group but not with the hydroxyl.

If the clause beginning with "that" is deleted, the sentence does not convey the information intended. Therefore, the clause is restrictive.

When a clause is nonrestrictive, it adds information, but the sentence does not lose its meaning if the clause is deleted. Nonrestrictive clauses are introduced by "which" and are set off by commas.

> The current-voltage curves, which are shown in Figure 6, clearly demonstrate the reversibility of all four processes.

> Several hazardous waste disposal sites are located along the shores of the Niagara River, which is a major water source.

Dangling Modifiers

One form of a dangling modifier is a verbal phrase that does not refer clearly and logically to another word or phrase in the sentence.

> *Incorrect*
> Understanding the effect of substituents on the parent molecules, the ortho hydrogens could be assigned to the high-frequency peak.

> *Correct*
> Understanding the effect of substituents on the parent molecules, we could assign the ortho hydrogens to the high-frequency peak.

> *Incorrect*
> Using the procedure described previously, the partition function can be evaluated.

> *Correct*
> The partition function can be evaluated by using the procedure described previously.

> *Also correct*
> Using the procedure described previously, we can evaluate the partition function.

> *Incorrect*
> When sprayed with ninhydrin reagent, no amino acids were detected.

> *Correct*
> When the chromatogram was sprayed with ninhydrin reagent, no amino acids were detected.

Awkward Omissions of Verbs and Auxiliary Verbs

Be sure that each subject in a compound sentence has the proper verbs and auxiliary verbs.

Incorrect
The eluant was added to the column, and the samples collected in 10-mL increments.

Correct
The eluant was added to the column, and the samples were collected in 10-mL increments.

Punctuation

Comma

- Use a comma before and after Jr. and Sr., but not necessarily II or III.

- Use a comma in a date after the day, but not after the month when the day is not given.

    June 15, 1984   *or*   June 1984

- Within a sentence, use a comma after the year as well.

    On August 18, 1984, an important event took place.

- Do not use a comma after the subordinating conjunction in a nonrestrictive phrase.

    Incorrect
    The bryopyran ring system is a unique requirement for anticancer activity whereas, the ester substituents influence the degree of cytotoxicity.

    Correct
    The bryopyran ring system is a unique requirement for anticancer activity, whereas the ester substituents influence the degree of cytotoxicity.

- Use a comma after a long introductory clause or phrase.

    Because of the known reactivity of α-chloro sulfides, it is not surprising that compound **10** is easily converted to the α-hydroxy by water.

- Use commas both before and after nonrestrictive phrases or clauses.

  The products, which were produced at high temperatures, were unstable.

- In series of words or phrases containing three or more items, use commas before "and" and "or".

  water, sodium hydroxide, and ammonia

  The red needles were collected, washed with toluene, and dried in a vacuum desiccator.

- Use commas to cite two references, but use an en dash (typed as a hyphen) for three or more in sequence, whether they are superscript numbers or on the line in parentheses.

  Certain complexes of cobalt were reported (*10, 11*) to have catalytic effects on hydrolysis reactions.

  Certain complexes of cobalt were reported (*10–12*) to have catalytic effects on hydrolysis reactions.

  Experimental investigations[10,14,18-25] concerned the relative importance of field and electronegativity effects.

- Use a comma before the coordinating conjunctions "and", "or", "nor", "but", "yet", "for", and "so" connecting two main clauses (complete thoughts).

  Toluene and hexane were purified by standard procedures, and benzene was redistilled from calcium hydride.

  No dielectric constants are available for concentrated acids, so it is difficult to give a quantitative explanation for the results.

  Exception: If both clauses are very short, the comma may be omitted.

- Do not use a comma in a sentence with a compound predicate that is not a member of a series.

  Incorrect
  The product distribution results were obtained in sodium hydroxide, and are listed in Table X.

  Correct
  The product distribution results were obtained in sodium hydroxide and are listed in Table X.

- Use commas to set off "that is", "namely", and "for example". In parenthetical expressions, use a comma after "i.e." and "e.g."

- Do not use a comma preceding "et al." except in a series.

    Jones et al.   Brown, Smith, et al.

- Do not use a comma to separate a verb from its subject or its object.

    Incorrect
    The addition of substituted silanes to carbon–carbon double bonds, has been studied extensively.

    Correct
    The addition of substituted silanes to carbon–carbon double bonds has been studied extensively.

    Incorrect
    The solvents employed in this study were, cyclohexane, methanol, *n*-pentane, and toluene.

    Correct
    The solvents employed in this study were cyclohexane, methanol, *n*-pentane, and toluene.

- Use a comma to introduce direct questions or quotations, but not if the question or quotation is the subject or object of a sentence.

    In the words of Pasteur, "Chance favors the prepared mind."

    Pasteur said "Chance favors the prepared mind."

    "Chance favors the prepared mind" is a translation from the French.

- In chemical names, use a comma between numerical locants, chemical element symbol locants, and Greek locants, without a space.

    1,2-dinitrobutane   *N*,*N*-dimethylacetamide

    β,4-dichlorocyclohexanepropionic acid

Quotation Marks

Location of quotation marks is a style point in which ACS differs from other authorities. In 1978, ACS questioned the practice and recommended a deviation from it: logical placement. Thus, if the punctuation is part of the quotation, then it should be within the quotation marks; if the punctuation is not part of the quotation, the writer should not mislead the reader by inferring that it is.

- Place quotation marks before all punctuation that is not part of the original quotation. Place them after all punctuation that is part of the quotation.

   The sample solution was stirred briefly with a magnetic "flea".

   Ralph Waldo Emerson said "The reward of a thing well done is to have done it."

- Use quotation marks for new words, words used in a new sense, or words not used literally, but only the first time they appear in your text.

   Plastocyanin is a soluble "blue" copper protein.

   The integrated intensity of each diagonal in the spectrum is proportional to a "mixing coefficient".

   The so-called "electron-deficient" cations are, in fact, well-established intermediates.

- Use quotation marks to enclose short direct quotations (two or three sentences).

   In the book *Megatrends,* Naisbitt concludes that "We are moving from the specialist who is soon obsolete to the generalist who can adapt."

- Use a narrower page width and no quotation marks for longer quotations (extracts) of 50 words or more.

   Everything is made of atoms. That is the key hypothesis. The most important hypothesis in all of biology, for example, is that everything that animals do, atoms do. In other words, there is nothing that living things do that cannot be understood from the point of view that they are made of atoms acting according to the laws of physics.

   —Richard Phillips Feynman

- Use single quotation marks only when they are within double quotation marks.

   He said "I consulted Beilstein's 'Handbuch' to verify this information."

Parentheses

In correct usage, parenthetical expressions contain information that is subsidiary to the point that the sentence is making. The sentence does not depend on the information within the parentheses.

- Use parentheses for parenthetical expressions that clarify, identify, or illustrate and that direct the reader.

> The total amount (10 mg) was recovered by modifying the procedure.
>
> The final step (washing) also was performed under a hood.
>
> The curve (Figure 2) obeys the Beer–Lambert law.
>
> The results (see Table I) were consistently positive.
>
> Only 15 samples (or 20%) were analyzed.

- Use parentheses to enumerate. Use parentheses in pairs, not singly.

> *Incorrect*
> Three applications of this reaction are possible: 1) isomerization of sterically hindered aryl radicals, 2) enol–keto transformation, and 3) sigmatropic hydrogen shift.
>
> *Correct*
> Three applications of this reaction are possible: (1) isomerization of sterically hindered aryl radicals, (2) enol–keto transformation, and (3) sigmatropic hydrogen shift.

- If items are numbered for future citing in text, do not put the numbers in parentheses. Such numbers are not parenthetical.

> Two theories are currently used to explain this behavior: 1, the wettability factor; and 2, the adsorption influence. Theory 1 postulates that wettability is inversely proportional to strength.

- Do not use parentheses when citing a reference or equation in narrative. In such a case, the reference or equation number is the point of the sentence, not subsidiary information, and thus not parenthetical.

> in ref 12   *not*   in ref (12)
>
> as shown in eq 10   *not*   as shown in eq (10)

- Use parentheses to identify the trademark and manufacturer of reagents and equipment.

> cobalt chloride (Mallinckrodt)
>
> a pH meter with a glass electrode (Corning)

- If a parenthetical sentence is within another sentence, do not use a final period.

  Our results (see Figure 5) justified our conclusions.

- If a parenthetical sentence is not within another sentence, use a final period inside the closing parenthesis.

  A mechanism involving loss of a CH radical followed by rearrangement was proposed. (See Scheme I.)

- Enclose oxidation numbers in parentheses, closed up to the element name or symbol, in text but not in formulas. (In formulas use superscripts.)

  copper(III)   *or*   Cu(III)

  iron(II)   *or*   Fe(II)

- Use parentheses closed up to the compound names to indicate isotopic substitution.

  (^{15}N)ammonia (^{14}C)glucose

  (Square brackets are used for isotopic labeling.)

Colon

- Use a colon to introduce a word, phrase, complete sentence, or several complete sentences that illustrate, clarify, or expand the information that precedes it.

  The electron density was studied for the ground state of three groups of molecules: (1) methane–methanol–carbon dioxide, (2) water–hydrogen peroxide, and (3) ferrous oxide–ferric oxide.

  We now report a preliminary finding: No chemical shift changes were detected in the concentration range 0.1–10 M.

  The following are our conclusions: Large-angle X-ray scattering studies give us an accurate picture of structures up to 9 Å. They do not allow the specification of defects, such as random ruptures of the chains. The structural models defined are strongly supported by magnetic measurements.

- Use a colon to express numerical ratios. A slant (solidus) is also acceptable.

  10:1   *or*   10/1

- Do not use a colon (or any punctuation) between a verb and its object or a preposition and its object.

    Incorrect
    The rate constants for the reaction in increasing concentrations of sodium hydroxide are: 3.9, 4.1, 4.4, 4.6, and 4.9.

    Correct
    The rate constants for the reaction in increasing concentrations of sodium hydroxide are 3.9, 4.1, 4.4, 4.6, and 4.9.

Dashes

The shortest dash is the hyphen (-); the en dash (–) is longer; and the em dash (—) is the longest. On a typewriter, use one hyphen for hyphens and en dashes; use two hyphens for em dashes. A copy editor will mark en and em dashes for the typesetter.

Hyphens

Hyphens are discussed in the section on "Hyphenation".

En Dash

- Use an en dash to mean the equivalent of "and" or "to" in two-word concepts.

structure–activity relationships	producer–user communication
cis–trans isomerization	nickel–cadmium battery
oxidation–reduction potential	ethanol–ether mixture
dose–response relationship	host–guest complexation
carbon–oxygen bond	metal–metal bonding
bromine–olefin complex	temperature–time curve
vapor–liquid equilibrium	metal–ligand complex

- Use an en dash to mean "to" or "through" with a span of three or more numerals.

    12–20 months   Figures 1–4   5–50 kg

- Do not use an en dash when the words "from" or "between" are used.

    from 500 to 600 mL   between 7 and 10 days

- Use an en dash to link two names.

Friedel–Crafts reaction	Diels–Alder reaction
Jahn–Teller effect	Debye–Hückel theory
Stern–Volmer plot	Beer–Lambert law
van't Hoff–Le Bel	Michaelis–Menten kinetics
Lineweaver–Burk	

- Use an en dash between components of a mixed solvent.

  The melting point was unchanged after three crystallizations from hexane-benzene.

  (A solidus may also be used.)

Em Dash

- Use em dashes to set off words that would be otherwise misread.

  Incorrect
  All three experimental parameters, temperature, time, and concentration were strictly followed.

  Correct
  All three experimental parameters—temperature, time, and concentration—were strictly followed.

- Do not use em dashes to separate phrases or nonrestrictive clauses if another form of punctuation can be used.

  Incorrect
  Jones—not Riesberg—obtained good correlation of results and calculations.

  Correct
  Jones, not Riesberg, obtained good correlation of results and calculations.

  Incorrect
  The singly charged complexes—which constituted bands 1 and 3—liberated maleate anion upon decomposition.

  Correct
  The singly charged complexes, which constituted bands 1 and 3, liberated maleate anion upon decomposition.

Solidus (or Slash)

- Use a solidus for simple fractions and in all subscript and superscript fractions.

  a/b $(x + y)/(3x - y)$ $x^{1/2}$

- Use a solidus as a symbol for "per" in abbreviated units of measure.

    10 kg/cm^2 100 m/s

- Do not use a solidus between words in text. Use of the solidus to mean "and" or "or" is not promoted by authoritative style guides. For example, in "and/or", the solidus means "or", but this combination is much overused. Closer reading of the text often reveals that either "and" or "or" is sufficient. Furthermore, an en dash is the preferred punctuation to mean "and".

    Exception: A solidus may be used between components of a mixed solvent.

    hexane/benzene

Period

- Do not use periods after abbreviated units of measure and most other abbreviations and symbols, except when the abbreviation could be confused with another word (in. for inches, at. for atomic, no. for number).

    Exceptions

A.D.	anno Domini
a.m.	ante meridiem
anal.	analysis
ca.	circa
cf.	compare
ed.	edited, edition
Ed.	editor
e.g.	for example
ibid.	in the same place
i.d.	inside diameter
i.e.	that is
o.d.	outside diameter
p.m.	post meridiem
v.	versus, in legal citations
vs.	versus

- Use periods between numbers within square brackets to form names for bridged and spiro alicyclic compounds.

    bicyclo[3.2.0]heptane   spiro[4.5]decane

Ellipsis Points

- Within a quotation, use three periods (points of ellipsis) to indicate deleted words or phrases. These three periods are in addition to other needed punctuation. Thus, if a period is already there, the result will be four periods.

   > No science is immune to the infection of politics and the corruption of power.... The time has come to consider how we might bring about a separation, as complete as possible, between Science and Government in all countries.
   > —*Jacob Bronowski*

- Generally, do not begin or end a quotation with ellipsis points.

Square Brackets

- Use square brackets within quotation marks to indicate material that is not part of a direct quote.

   > In the words of Sir William Lawrence Bragg, "The important thing in science is not so much to obtain new facts as to *discover new ways* [italics added] of thinking about them."

- Use square brackets to indicate concentration: [Ca].

- Use italic letters within square brackets to form names for polycyclic aromatic compounds.

   dibenzo[*c*,*g*]phenanthrene   dibenz[*a*,*j*]anthracene
   dicyclobuta[*de*,*ij*]naphthalene   indeno[1,2-*a*]indene
   1*H*-benzo[*de*]naphthacene

- Use numbers separated by periods within square brackets to form names of bridged and spiro alicyclic compounds.

   bicyclo[3.2.0]heptane   spiro[4.5]decane

- Use square brackets closed up to the compound name to indicate isotopic labeling.

   [^{15}N]alanine   [2-^{14}C]leucine

(Parentheses are used for isotopic substitution.)

Semicolon

- Use a semicolon to separate independent clauses if no conjunction is used.

  Incorrect
  All solvents were distilled from an appropriate drying agent, tetrahydrofuran and diethyl ether were also pretreated with activity I alumina.

  Correct
  All solvents were distilled from an appropriate drying agent; tetrahydrofuran and diethyl ether were also pretreated with activity I alumina.

- Use a semicolon between independent clauses joined by conjunctive adverbs such as "that is", "however", "therefore", and "thus".

  Incorrect
  The proposed intermediate is not easily accessible, therefore, the final product is observed initially.

  Correct
  The proposed intermediate is not easily accessible; therefore, the final product is observed initially.

- Do not use a semicolon between dependent and independent clauses.

  Incorrect
  The activity on bromopyruvate was decreased; whereas, the activity on pyruvate was enhanced.

  Correct
  The activity on bromopyruvate was decreased, whereas the activity on pyruvate was enhanced.

- Use semicolons between items in a series if one or more of the items already contain commas.

  Persons in attendance were James Taven, University of Maryland; Anne Schmidt, MIT; and Robert Berren, The Ohio State University.

  This rule holds even if the only group containing the commas is the last in the series.

  The compounds studied were methyl ethyl ketone; sodium benzoate; and acetic, benzoic, and cinnamic acids.

Spelling

- Use a dictionary. *Webster's New World Dictionary of the American Language* is the desk dictionary used by the ACS technical editing staff, who will change spelling variants to the form recommended therein. The unabridged dictionary that the ACS staff uses is *Webster's Third New International Dictionary*. However, whatever dictionary you have, you should usually choose the first spelling and original meaning of a word, as opposed to the fourth or fifth definition.

- Many words in regular usage, as well as many technical terms, have two or more acceptable spellings. The following list gives recommended spellings and capitalizations where appropriate; included are some terms not found in easily accessible dictionaries, words often misspelled, and common expressions.

absorbance
absorbency
aerobic
aging
aglycon
air-dry (verb)
amine (RNH_2)
ammine (NH_3 complex)
ampule
analogue
antioxidant
aqua regia
Arrhenius
artifact
asymmetry
audio frequency
autoxidation
Avogadro
back-donation
backscattering
back-titrate (verb)
bandwidth
base line
Beckman (instrument company)
Beckmann (thermometer, rearrangement)

Beilstein
bit
blackbody
blackbox
Bragg scattering
break-seal
Büchner
buildup (noun)
build up (verb)
buret
butanol, 1-butanol (not *n*-butanol)
tert-butylation
bypass
byproduct
byte
canceled
Cartesian
catalog
clear-cut
coauthor
co-ion
collinear
condensable
conductometric
conrotatory
coordination
Coulombic

counter electrode
counterion
co-worker
cross-link
cross section
cuboctahedron
cuvette
data base
deamino
deoxy
desiccator
deuterated
deuterio
deuterioxide
deuteroporphyrin
Dewar benzene
dialogue
disc (electrophoresis)
discernible
disk
disrotatory
dissymmetric
distill
drybox
dry ice
eigenfunction
eigenvalue
electronvolt

electron microscope
eluant
eluate
eluent
enzymatic
Erlenmeyer
faradaic
far-infrared
filterable
flavin
formulas
forward
freeze-dry (verb)
fulfill
Gaussian
gegenion
glovebag
glovebox
graduated cylinder
gray
half-life
Hamiltonian
heat-treat (verb)
hemoglobin
hemolysate
heterogeneous
homogeneous
homologue
hydrindan
hydriodide
hydrolysate
hydrolyzed
indan
indexes
indices (crystallography)
inflection
in vacuo
isooctane
isosbestic
Kekulé
Kjeldahl
labeled
least squares (noun)
leukocyte
leveling

levorotatory
lifetime
ligancy
line width
liquefy
lysed
makeup (noun)
Markovnikov
metalation
methyl Cellosolve
methyl orange
midpoint
Millipore
mixture melting point
Mössbauer
near-ultraviolet
ortho ester
orthoformate
outgassing
overall
parametrization
path length
percent
Petri
pharmacopeia
phlorin
phosphomonoester
phosphorus
pipet
2-propanol (not
    isopropanol)
pseudo first order
    (noun)
pseudo-first-order
    (adjective)
pyrolysate
quantitation
radioelement
radio frequency
radioiodine
reexamine
reform (to amend)
re-form (to form again)
riboflavin
ring-expand (verb)

rotamer
round-bottomed or
    round-bottom
sideband
side chain
Soxhlet
spin-label (noun)
steam-distill (verb)
stereopair
superacid
superhigh frequency
supernate (noun)
supernatant (adjective)
sulfur
syndet
syrup
test tube
thermostated
theta solvent
thiamin
thio acid
thioether
toward
transmetalation
tropin
Ubbelohde
ultra-high-vacuum
un-ionized
uni-univalent
upfield
urethane
VandenHeuvel
van der Waals
van't Hoff–Le Bel
wave function
wavelength
wavenumber
well-known
work up (verb)
workup (noun)
X irradiation
X-ray
ylide

- Use American spellings, except in names and direct quotations.

- Form the possessive of a joint owner by adding an apostrophe and an "s" after the last name only.

  Kanter and Marshall's results   Bausch and Lomb's equipment

- Form the possessive by adding an apostrophe and an "s" to singular nouns, including proper nouns already ending in "s".

  Burns's poems

Editorial Style

Hyphenation

- Do not hyphenate most prefixes added to common nouns, even if a double letter will result. The following prefixes usually are not hyphenated when added to common nouns.

after	extra	non	stereo
ante	hyper	over	super
anti	hypo	photo	supra
auto	infra	physico	trans
bi	iso	poly	tri
bio	metallo	post	ultra
co	mid	pre	un
counter	macro	pro	under
de	micro	pseudo	up
di	mini	re	visco
down	mono	semi	
electro	multi	sub	

Examples
precooled	*not*	pre-cooled
multicolored	*not*	multi-colored
nonpolar	*not*	non-polar
antibacterial	*not*	anti-bacterial
microorganism	*not*	micro-organism
cooperation	*not*	co-operation

Exceptions
anti-infective	*not*	antiinfective
co-worker	*not*	coworker
un-ionize	*not*	unionize
co-ion	*not*	coion

- Do not hyphenate a common noun and the suffix "like" unless a triple "el" will result.

  catlike   *but*   doll-like

- Do not hyphenate a common noun and the suffix "fold".

  fivefold   multifold

- Do hyphenate the suffix "fold" to a number.

  25-fold

- Do not hyphenate a common noun and the suffix "wide".

  worldwide   statewide

- If a prefix is added to a two-word compound and refers to the compound, hyphenate it.

  non-radiation-caused effects   pseudo-first-order reaction
  non-diffusion-controlled system   pre-steady-state condition

- Hyphenate prefixes and suffixes added to proper nouns, and retain the capital letter.

  non-Gaussian   non-Newtonian   Kennedy-like

- In chemical names, use hyphens to separate locants and configurational descriptors from the syllabic portion of the name. Locants and descriptors can be numbers, element symbols, small capital letters, Greek letters, and italic words and letters. Do not place a hyphen in a chemical name if it is not separating a locant or descriptor from the syllabic portion. Do not use a hyphen to separate the syllables of a chemical name unless the name is too long to fit on one line. See the section "Recommended End-of-Line Hyphenation of Chemical Names".

  2-benzoylbenzoic acid   *N*-methylmethanamine
  α-ketoglutaric acid   3-chloro-4-methylbenzoic acid
  toluene-d_6 *cis*-dichloroethylene
  D-arabinose   *trans*-2-bromocyclopentanol

- Hyphenate a prefix to a chemical name.

    non-hydrogen bonding   non-phenyl atoms

Compound Words and Unit Modifiers

Compound words and unit modifiers are hyphenated. *Compound words* are two or more terms used to express a single idea (examples: cross-link, son-in-law). Compound words in common usage are listed in most dictionaries. *Unit modifiers* are two words used as an adjective; they may consist of a noun and an adjective (e.g., time-dependent reaction, radiation-sensitive compound, water-soluble polymer, halogen-free oscillator), an adjective and a noun (e.g., high-frequency transition, small-volume method, first-order reaction, outer-sphere redox couple), an adverb and an adjective (e.g., above-average results, still-unproven technique), or two nouns (e.g., ion-exchange resin, liquid-crystal polymers, transition-state modeling, charge-transfer reaction, gas-phase hydrolysis). A short list of compound words and unit modifiers commonly seen in ACS publications is at the end of this section. Also, the *GPO Style Guide* has an excellent guide to compounding, which ACS technical editors follow.

- Do not hyphenate unit modifiers if the first word is an adverb ending in "-ly".

    recently developed procedure   carefully planned experiment
    accurately measured values   poorly written report

- When a number and a unit of time or measure are used as an adjective, hyphenate them.

    a 12-min exposure   a 20-mL aliquot   a 10-mg sample

    Exceptions

    a 37 °C water bath   a 0.1 M NaOH solution

If the unit of measure is complex, do not use a hyphen.

    a 0.1 mol dm^{-3} solution

- When converted units are given in parentheses and are also adjectival, hyphenate them.

    a 7-in. (17.8-cm) funnel   a 0.25-in. (6.4-mm) conductor

- When an adjectival unit modifier contains an en dash between numbers, also hyphenate it between the last number and the unit of measure.

    a 1–2-h sampling time   a 25–30-mL aliquot

- When two or more unit modifiers have the same base, use a hyphen after each element, and do not repeat the base.

    100-, 200-, and 300-mL aliquots

    25- to 50-mg samples

    0.5- × 10-cm tube   *but*   5 × 5 cm² tube

    high-, medium-, and low-frequency measurements

    first- and second-order reactions

- Do not hyphenate very familiar unit modifiers or those that are naturally and easily recognizable.

    stainless steel flask   melting point determination
    molecular weight determination   data acquisition systems
    high molecular weight enzyme

- In enzyme names, hyphens may be used by IUB and not by Chemical Abstracts Service to connect chemical names in an adjectival sense.

    IUB: ribitol-5-phosphate dehydrogenase (EC 1.1.1.137)

    CA: ribitol 5-phosphate dehydrogenase (E.C. 1.1.1.137)

    Use either system, but be consistent throughout the paper.

- Do not hyphenate foreign phrases used as adjectives.

    in vivo reactions   ad hoc committee   in situ evaluation

- Hyphenate unit modifiers containing adverbs such as "well", "still", or "ever".

    well-known scientist   ever-present danger   still-new equipment

    Do not hyphenate these words if they are modified by another adverb.

    very well studied hypothesis

● Do not hyphenate unit modifiers that are chemical names.

> amino acid level   barium sulfate precipitate
> sodium hydroxide solution

● Do not hyphenate unit modifiers if one of the words is a proper name.

> Lewis acid catalysis   Schiff base measurement
> Fourier transform technique

● Do not hyphenate unit modifiers if the first word is a comparative or superlative.

> higher temperature reactions   lowest frequency wavelengths

Exceptions

> least-squares analysis   nearest-neighbor interaction

● Hyphenate unit modifiers that contain numbers.

> three-dimensional model   one-electron transfer
> two-phase system   two-compartment model
> five-coordinate complex   seven-membered cyclic
> three-stage sampler   derivative
> three-neck flask

● Hyphenate unit modifiers that contain a verb or a present or past participle.

> methyl-substituted intermediate   immobilized-phase method
> air-equilibrated samples   fluorescence-quenching solution
> rate-limiting step   laser-induced species
> ion-promoted reaction   problem-solving abilities

● Do not hyphenate unit modifiers containing three or more words, even if similar two-word modifiers are hyphenated, when doing so would contravene other recommendations.

> Lewis acid catalyzed reactions
>
> copper-to-iron ratio   *but*   sodium chloride to iron ratio

- Hyphenate unit modifiers containing three words or a number, unit of measure, and a word.

  out-of-plane distance   signal-to-noise ratio
  3-year-old child   4-mm-thick layer

- Hyphenate unit modifiers containing three words when similar two-word modifiers are hyphenated.

  acid-catalyzed reaction   general-acid-catalyzed reaction

  metal-promoted reaction   transition-metal-promoted reaction

- Hyphenate unit modifiers used as predicate adjectives. Usually, only unit modifiers that consist of nouns and adjectives can be used as predicate adjectives.

  In these cluster reactions, dehydrogenation is size-dependent.

  All compounds were light-sensitive and were stored in the dark.

  The reaction is first-order.

  The complex is square-planar.

- The following is a list (by no means complete) of unit modifiers seen frequently. **These should be hyphenated when modifying a noun.**

air-dried
air-equilibrated
back-bonding
charge-transfer
diffusion-controlled
electron-diffraction
electron-transfer
energy-transfer
first-order
flame-ionization
fluorescence-quenching
free-radical
gas-phase
Gram-positive
halogen-free
high (low)-energy
high (low)-frequency
high-performance
high (low)-pressure
high (low)-resolution

high (low)-temperature
ion-exchange
ion-promoted
ion-selective
^{14}C-labeled
laser-induced
least-squares
light-catalyzed
long (short)-chain
long (short)-lived
nearest-neighbor
oil-soluble
outer-sphere
pseudo-first-order
radiation-caused
radiation-produced
radiation-sensitive
rate-limiting
reversed-phase
round-bottom

side-chain
short (long)-chain
short (long)-lived
small (large)-volume
steady-state
temperature-dependent
thin-layer
three-phase
time-dependent
transition-metal
transition-state
two-dimensional
water gas shift
  (not hyphenated)
water-soluble
water-insoluble
weak-field

Capitalization

- Generally, in text, keep everything lower case except proper nouns. However, there are many exceptions.

- In titles, headings, or names that are capital and lower case, do not capitalize coordinating conjunctions ("and", "but", "or", "nor", "yet", "so"), articles ("a", "an", "the"), or prepositions. Do capitalize other parts of speech, regardless of the number of letters. Do capitalize the "to" in infinitives. Do capitalize the first and last words of the title or heading, regardless of part of speech, unless the word is mandated to be lower case (e.g., pH).

   Reactions of Catalyst Precursors with Hydrogen and Deuterium

   Scope of the Investigations: The First Phase

   Properties of Organometallic Fragments in the Gas Phase

   Compounds To Be Tested

- Do not capitalize the "r" in X-ray at the beginning of a sentence or in titles and headings.

- Capitalize the words "figure", "table", "chart", and "scheme" only when they refer to a specific numbered item.

   Figure 1   Chart IV   Table II   Schemes IV–VII

- Do not capitalize "reference", "equation", or "structure", even when they refer to a specific numbered item.

   reference 3   structure **1**

- Capitalize parts of a book when they refer to a specific number

   Chapter 3   Appendix I   *but*   the preface   the contents

- Do not capitalize "page" even with a number.

   The photographs on page 3 . . .

- Capitalize only the name of the eponym but not the noun.

Avogadro's number	Einstein's theory	nuclear Overhauser
Hodgkin's disease	Lewis acid	effects
Raman spectroscopy	Graham's law	Schiff base

- Capitalize adjectives formed from proper names.

  Gaussian   Newtonian   Mendelian   Cartesian

  Exception

  faradaic

- Do not capitalize the first word after a colon if it is not the first word of a complete sentence.

  Two types of asymmetric reactions were conducted: synthesis of styrene oxide and reduction of olefinic ketones.

- Do not capitalize lower-cased abbreviations and symbols when they are at the beginning of a sentence or in a capitalized title.

  pH–shift relationships were studied.

  o-Dichlorobenzene was the solvent.

  Reaction of *trans*-4-(Phenylsulfonyl)-3-buten-2-one

- Use a capital letter as the subscript in describing certain reactions.

  S_N1 S_N2

- Always capitalize genus names, but never capitalize species names, even when they are in a capitalized title.

  Novel Metabolites of *Siphonaria pectinata*

  Bacillus subtilis   *Pneumococcus aureus*

- Do not capitalize the adjectival or plural form of a genus name unless it is at the beginning of a sentence or in a title.

  bacilli   pneumococcal   streptococcal

- In titles and headings with compound words, capitalize both words if the compound is a unit modifier.

  High-Temperature System   Base-Catalyzed Cyclization
  Deuterium-Labeling Experiment   Thyrotropin-Releasing Hormone

- Capitalize "Earth" as a planet.

- Capitalize each component of hyphenated words if the component would be capitalized when standing alone.

   Non-Hydrogen-Bonding   Quasi-Elastic   Zig-Zag   Ex-Senator
   Build-Up   Cross-Linked

- Do not capitalize chemical names and nonproprietary drug names unless they are the first word of a sentence or in a title or heading. In such cases, capitalize the first letter of the English word, not the locant, stereoisomer descriptor, or positional prefix. (See the section on "Chemical Names".)

- Polymer names are often two words in parentheses following the prefix "poly". In text they are lower case. As the first word of a sentence and in titles or headings, capitalize only the first letter ("P").

   Poly(vinyl chloride) is a less useful polymer than poly(ethylene glycol).

   Reactions of Poly(methyl methacrylate)

   New Uses for Poly(ethylene terephthalate)

- Capitalize trade names.

   Chow   Plexiglas

   In general, however, generic names are preferred to trade names.

   petroleum jelly   *not*   Vaseline

   photocopy   *not*   Xerox

   borosilicate glass   *not*   Pyrex

   mineral oil   *not*   Nujol

   poly(tetrafluoroethylene)   *not*   Teflon

- Do not capitalize the common names of equipment.

spectrophotometer	mass spectrometer
temperature controller unit	electron-diffraction chamber
mercury lamp	flame-ionization detector
gas chromatograph	

- In titles and headings, do not capitalize abbreviated units of measure; do capitalize spelled-out units of measure.

   Analysis of Milligram Amounts   *but*   ...of 2 mg

- Capitalize sections of the country but not the corresponding adjectives.

    Northeast   *but*   northeastern

    Midwest   *but*   midwestern

- Although a current trend is to lower case the names of persons when these names are used as adjectives and the persons have become very familiar, many are still capitalized. The following is a list (by no means complete) of names that should be capitalized.

Avogadro	Coulombic	Markovnikov
Beckmann (thermometer, rearrangement)	Dewar benzene	Mössbauer
	Erlenmeyer	Petri
Beilstein	Gaussian	Scatchard
Bragg scattering	Gram	VandenHeuvel
Büchner	Hamiltonian	van der Waals
Bunsen	Kekulé	van't Hoff
Cartesian	Kjeldahl	

Surnames and Formal Names

Most people are aware that the Chinese use their surnames first, then their given names. For example, Sun Yat-sen is Dr. Sun. However, the problem of identifying surnames extends to many other cultures. This multiplicity of usage creates problems in bibliographic indexes and in reference citations. The reference citation should always list the surname first, followed by initials. The authors of papers should be listed in standard American format to ensure greater consistency of citation practice. If a footnote would clarify the situation or eliminate any perceived confusion, use a footnote.

   In most cultures, the surname is the family name, but it may not be the formal name, that is, the name or shortest string of names that can be used following a title. The following are some cases in which the surnames are not the formal names. This list is by no means complete, but at least it will help you to be aware of these differences.

Spanish—Frequently three or more names; the last two are surnames, sometimes connected by "y". The second surname is often dropped or abbreviated. The formal name begins with the first surname and includes the second surname only in very formal usage. Example: Juan Perez Avelar is Dr. Perez Avelar.

Hungarian—Two names; the surname is first, and it is the formal name. However, the second name is accepted as formal internationally.

Arabic—Often many names, and the position of the surname is highly variable. The formal name often consists of two or three names including articles that can be joined.

Thai—Two names; the surname is last, but the formal name is first.

Vietnamese—Two or three names. The first is the surname and formal name.

Japanese—Two names; the surname is the formal name. The surname is first in Japanese. However, when they translate their names into non-Asian languages, they place their surnames last. Example: Taro Yamada is Dr. Yamada.

Korean—Usually three names; the surname is first and is the formal name. In non-Asian usage, they sometimes place the surname last.

Numeral Usage

- Use consecutive numerals for figures, tables, schemes, structures, and references. Use Arabic numbers for references, figures, and structures. Use Roman numbers for tables and schemes. No publication allows skipping numbers or numbering out of sequence.

- Use numerals with units of time or measure, and use a space between the number and the unit (except %).

    6 min   25 mL   125 mg/kg   0.30 g   50% $250

    50 mL of water/g of compound   2 half-lives

    Exception: in more complex situations

    50 mL of water and 20 mg of NaOH per gram of compound

- When measurements are used in a nontechnical sense in nontechnical text, spell them out.

    If you take five minutes to read this article, you'll be surprised.

- With items other than units of time or measure, use words for numbers less than 10; use numerals for 10 and above, except as the first word of a sentence.

    three flasks   *but*   30 flasks

    seven trees   *but*   10 trees

- Numbers may be used to name an item.

    Sample 1 contained a high level of contamination, but samples 2 and 3 were found to be relatively pure.

- Numbers are used in a mathematical sense.

  The incidence of disease increased by a factor of 4.

  The yield of product was decreased by 6 orders of magnitude.

- For very large numbers, in narrative text, use a combination of numerals and words.

  1 billion tons   4.5 billion years
  180 million people $15 million (*not* 15 million
  2 million pounds (*not* lb)   dollars)

- For very large numbers in scientific notation, use exponents.

  $t_{1/2} = 1.2 \times 10^6$ s

- Use all numerals in a series containing numbers 10 or greater.

  5, 8, and 12 experiments

- Try not to start a sentence with a numeral. Recast the sentence if possible, but if not, spell out the number and the unit of measure if there is one.

  Twelve species were evaluated in this study.

  Twenty slides of each blood sample were prepared.

  Fifteen milliliters of supernate was added to the reaction vessel.

- Even when a sentence starts with a spelled-out number, follow the rules for other numbers in the sentence.

  Three micrograms of sample was dissolved in 20 mL of acid.

  Fifty samples were collected, but only 22 were tested.

- Use numerals for decades, and form their plurals by adding an "s".

  the 1960s   the 1900s   *not*   the '50s

- When numbers are used as names and not to enumerate, form their plurals by adding an apostrophe and "s" to avoid confusion with mathematical expressions.

  Many 6's were registered.

- Use decimals rather than fractions with units of time or measure, except when doing so would imply an unwarranted accuracy.

   3.5 h   *not*   3½ h

   5.25 g   *not*   5¼ g

- In descriptive usage, spell out and hyphenate fractions.

   one-quarter of the experiments   two-thirds of the results

- Spell out ordinals "first" through "ninth"; use numerals for "10th" and greater.

- Do not add ordinal endings to numbers in dates.

   January 3   *not*   January 3rd

   September 5   *not*   September 5th

- For lists of phrases, use Arabic numerals in parentheses. Always use two parentheses, not one.

   Some advantages of these materials are (1) their electrical properties after pyrolysis, (2) their ability to be modified chemically before pyrolysis, and (3) their abundance and low cost.

- For lists of sentences, use Arabic numerals followed by periods when the sentences are displayed; use Arabic numerals in parentheses when the sentences are run into text.

   These results suggest the following:
   1. Ketones are more acidic than esters.
   2. Cyclic carboxylic acids are more acidic than their acyclic analogues.
   3. Alkylation of the active methylene carbon reduces the acidity.

   The major conclusions are the following: (1) We have further validated the utility of molecular mechanical methods in simulating the kinetics of these reactions. (2) A comparison of the calculated structures with available X-ray structures revealed satisfactory agreement. (3) The combined use of different theoretical approaches permitted characterization of the properties of a new isomer.

- When two numbered items are cited in narrative, use "and".

   Figures 1 and 2   references 23 and 24   Tables I and II

- Use an initial zero before a decimal: 0.25.

- Use a comma with reference callouts in parentheses or as superscripts.

    Lewis (*12, 13*) found that   Lewis[12,13] found that

    When the numbers are on the line, the comma is followed by a space; when the numbers are superscripts, the comma is not followed by a space. This style is strictly a typesetting convention.

- Use an en dash (typed as a hyphen) with three or more numbered items, both in narrative and in parentheses.

    Tables II–IV show that   Past results (*27–31*) indicate that
    References 3–5 are reviews of

- Use words and do not hyphenate adjectives formed with the suffix "fold" if the number is less than 10. Hyphenate such adjectives and use a numeral when the number is 10 or above and when the context is primarily mathematical rather than narrative.

    twofold   *but*   20-fold

- Use numbers with a.m. and p.m.

    12:15 a.m.   4:00 a.m.

- Use numbers when discussing positions in chemical structures.

    the 1-position of the ring

    the carbon in the 6-position   *or*   C-6

    not   carbon-6   (the hyphen is used for isotopic designation)

- To count atoms, use a subscript with the element symbol, but spell out the number if you spell out the element name.

    C_6 *or*   six carbons

    a six-carbon ring

Abbreviations and Acronyms

In an abbreviation, you pronounce the individual letters; in an acronym, the letters form a pronounceable word. ACS is an abbreviation; CASSI is an acronym. A list of abbreviations and acronyms follows this discussion.

- Use abbreviations sparingly. If a very long name or term is repeated many times throughout a paper, an abbreviation is warranted.

- Avoid abbreviations in the title of a paper.

- Place abbreviations in parentheses following the spelled-out forms the first time they appear in the text. If they are used in the abstract, define them in the abstract and again in the text.

- The following exceptions need not be defined at all:

at. wt	atomic weight
bp	boiling point
CP	chemically pure
ca.	circa (about)
cf.	confer (compare)
DNA	deoxyribonucleic acid
ed.	edited, edition
Ed.	Editor
equiv wt	equivalent weight
e.g.	for example
fp	freezing point
GLC	gas–liquid chromatography
IR	infrared
i.d.	inside diameter
mp	melting point
mmp	mixture melting point
M_r	molecular weight (relative molecular mass)
NMR	nuclear magnetic resonance
o.d.	outside diameter
RNA	ribonucleic acid
sp ht	specific heat
sp vol	specific volume
i.e.	that is
UV	ultraviolet
USP	United States Pharmacopeia
vol	volume
v/v	volume per volume
wt	weight
w/w	weight per weight

Furthermore, abbreviations that are common to a specific field may be permitted without identification in books and journals in that field only, at the discretion of the editor.

- Use "e.g." and "i.e." in figure captions, in tables, and in parentheses in text. If not in parentheses, spell out "for example" and "that is".

- Do not confuse abbreviations and mathematical symbols. An abbreviation is usually two letters or more; a mathematical symbol should generally be only one letter, possibly with a subscript or superscript. An abbreviation is used in narrative text but seldom appears in equations; a mathematical symbol must be used in equations and may also be used in text. For example, in text with no equations, PE for potential energy is acceptable, but in mathematical text and equations, use E_p. Abbreviations are typeset in Roman type; mathematical symbols are typeset in italic type.

- Symbols for the chemical elements are not treated as abbreviations or mathematical symbols. They need not be defined; they are typeset in Roman type.

- In text, abbreviate units of measure when they follow a number. Otherwise, spell them out.

   9 mg/kg *or* 9 mg kg^{-1} *but*   measured in milligrams per kilogram

   Exception: Units of measure may be abbreviated in parentheses with the definitions of variables directly following an equation.

   $L = D/P_O$

   where L is the distance between particles (cm), D is the particle density (g/cm^3), and P_O is the partial pressure of oxygen (kPa).

- In column headings of tables and in axis labels of figures, abbreviate units of measure.

- Do not abbreviate
   —the words "day", "week", "month", and "year"
   —days of the week
   —months not used with a date
   —titles not used with a name
   —states not used with a city

- Use a solidus or negative superscript only with abbreviated units of measure.

   55 g/L *or* 55 g L^{-1} *not*   55 grams/liter

   reported in meters per second   *not*   in meters/second

- Use the following abbreviations and spelled-out forms for months with a day or day and year in footnotes, tables, figure captions, bibliographies, and lists of literature cited. Without a specific day, spell out all the months.

Jan.	April	July	Oct.
Feb.	May	Aug.	Nov.
March	June	Sept.	Dec.

In text, spell out all months with or without a specific day.

- Use the abbreviation U.S. as an adjective only; spell out United States as the noun form.

  U.S. science policy   chemical industry in the United States

- Spell out and capitalize "company" and "corporation" as part of company names when they appear in an author's affiliation. Abbreviate them elsewhere in text. After the first mention, drop them entirely.

  "Dow Chemical Company" on first mention, "Dow" thereafter

- Check the list at the end of this discussion to find an ACS-recommended abbreviation. If there is none, you may use a special abbreviation (one that you devise) provided that (1) it is not identical with an abbreviation of a unit of measure, (2) it does not involve a drug with a generic name, (3) it will not be confused with the symbol of an element or a group, (4) it does not hamper the reader's understanding, and (5) you do not use too many abbreviations.

- Form the plurals of all-capital abbreviations by adding a lowercase "s" only, with no apostrophe.

  HOMOs   PCBs

- Do not use the following abbreviations at all.

CMR	^{13}C magnetic resonance (use ^{13}C NMR)
DMR	^{2}H magnetic resonance (use ^{2}H NMR)
FIR	far-infrared (use far-IR)
FTS	Fourier transform spectroscopy (use FT spectroscopy)
MIR	mid-infrared (use mid-IR)
NIR	near infrared (use near-IR)
PMR	proton or phosphorus magnetic resonance (use ^{1}H or ^{31}P NMR)

- Use the two-letter abbreviations for state names. (Do not abbreviate state names in contribution lines of ACS journals.)

United States

AL	Alabama	MT	Montana
AK	Alaska	NE	Nebraska
AZ	Arizona	NV	Nevada
AR	Arkansas	NH	New Hampshire
CA	California	NJ	New Jersey
CO	Colorado	NM	New Mexico
CT	Connecticut	NY	New York
DC	District of Columbia	NC	North Carolina
DE	Delaware	ND	North Dakota
FL	Florida	OH	Ohio
GA	Georgia	OK	Oklahoma
HI	Hawaii	OR	Oregon
ID	Idaho	PA	Pennsylvania
IL	Illinois	RI	Rhode Island
IN	Indiana	SC	South Carolina
IA	Iowa	SD	South Dakota
KS	Kansas	TN	Tennessee
KY	Kentucky	TX	Texas
LA	Louisiana	UT	Utah
ME	Maine	VT	Vermont
MD	Maryland	VA	Virginia
MA	Massachusetts	WA	Washington
MI	Michigan	WV	West Virginia
MN	Minnesota	WI	Wisconsin
MS	Mississippi	WY	Wyoming
MO	Missouri		

Canada

AB	Alberta	NS	Nova Scotia
BC	British Columbia	ON	Ontario
LB	Labrador	PE	Prince Edward Island
MB	Manitoba	PQ	Quebec
NB	New Brunswick	SK	Saskatchewan
NF	Newfoundland	UT	Yukon Territory
NT	Northwest Territories		

List of Abbreviations and Symbols

a	antisymmetric
	atto (10^{-18})
	axial [*use as* 2(a)-methyl in names]
a	*a* axis
	absorptivity
	axial chirality (as in (*aR*)-6,6′-dinitrodiphenic acid)
a_0	$a_0 = 0.52917$ Å = 1 Bohr radius
A	ampere
Å	angstrom
A	absorbance [$A = \log (1/T)$]
	anticlockwise
α	rotation, specific rotation
$[\alpha]^t_D$	specific rotation at temperature *t* and wavelength of sodium D line
$[\alpha]^t_\lambda$	specific rotation at temperature *t* and wavelength λ
AAS	atomic absorption spectroscopy
abs	absolute
ac	alternating current
Ac	acetyl (therefore OAc, acetate)
acac	acetylacetonate(-o) (as ligand in coordination nomenclature)
AcCh	acetylcholine
AcChE	acetylcholinesterase
ACS	ACS grade reagent, e.g. (ACS for American Chemical Society is generally written out)
ACTH	adrenocorticotropin
A.D.	anno Domini
Ado	adenosine
ADP	adenosine 5′-diphosphate
AES	atomic emission spectroscopy
	Auger electron spectroscopy
af	audio-frequency
AFS	atomic fluorescence spectroscopy
AGU	anhydroglucose units
Ah	ampere-hour
ala	alanyl in genetics
Ala	alanyl (ala in coordination nomenclature; A as a single-letter code); alanine
alt	alternating, as in poly(A-*alt*-B)
a.m.	ante meridiem
AM	amplitude modulation
AMP	adenosine 5′-monophosphate
amu	atomic mass unit (amu, reference to oxygen, is deprecated; u (reference to mass of ^{12}C) should be used; however, we do not always change amu to u or vice versa)
anal.	analysis (Anal. in combustion analyses presentations)
anhyd	anhydrous

Ans	ansyl
ansyl	8-anilino-1-naphthalenesulfonate
antilog	antilogarithm
AO	atomic orbital (ao in selective instances, e.g., supers and subs)
ap	antiperiplanar
AP	appearance potential
API	American Petroleum Institute
APS	appearance potential spectroscopy
aq	aqueous
AR	analytical reagent (e.g., AR grade)
Ara	arabinose
ara-C	cytidine, with arabinose rather than ribose (arabinocytidine, *not* cytosine arabinoside)
arb unit	arbitrary unit (clinical)
Arg	arginyl; arginine (R)
as	asymmetric (also *asym*)
AS	absorption spectroscopy
Asa	β-carboxyaspartic acid
ASIS	aromatic solvent-induced shift
Asn	asparaginyl, asparagine (N)
Asp	aspartyl, aspartic acid (D)
Asx	Asn or Asp
asym	asymmetric (also *as*)
ATCC	American Type Culture Collection
atm	atmosphere
atom %	atom percent
ATP	adenosine 5′-triphosphate
ATPase	adenosinetriphosphatase
ATR	attenuated total reflection
at. wt	atomic weight
au	atomic units
AU	absorbance units
AUFS	absorbance units full scale
av	average
b	barn (area) (10^{-24} cm^2)
	broad (spectral; br is preferred but circumstances may dictate b)
b	*b* axis
	block, as in poly(A-*b*-B)
B	buckingham (10^{-26} esu cm^2)
	ring (italic in steroid names)
B	boat (carbohydrate conformation)
β	stereochemical descriptor
bar	unit of pressure; unit and abbreviation are the same
bbl	barrel
bcc	body-centered cubic (also bccub)
bccub	body-centered cubic (also bcc)
BDH	British Drug House
BEHP	bis(2-ethylhexyl) phthalate

BET	Brunauer–Emmett–Teller (adsorption isotherm)
BeV	billion electronvolts (also GeV)
Bi	biot
b.i.d.	twice a day
biol	biological(ly)
bipy	2,2′-bipyridine, 2,2′-bipyridyl (*use* bpy)
	4,4′-bipyridine, 4,4′-bipyridyl (*use* bpy)
bis-Tris	(bis(2-hydroxyethyl)amino)tris(hydroxymethyl)methane (also Bistris, Bis-Tris)
BL	bioluminescence
BM	Bohr magneton (*use* μ_B)
BN	bond number
BO	Born–Oppenheimer
BOD	biological oxygen demand
bp	boiling point
bpy	2,2′-bipyridine, 2,2′-bipyridyl
	4,4′-bipyridine, 4,4′-bipyridyl
br	broad (spectral; also b)
BSA	bovine serum albumin
Btu	British thermal unit
bu	bushel
Bu	butyl
BWR	Benedict–Webb–Rubin equation
Bz	benzoyl (also bz)
Bzac	benzoylacetone
Bzl	benzyl (also bzl)
c	centered (crystal structure)
	centi (10^{-2})
	cyclo
c	*c* axis
	concentration, for rotation ($[\alpha]^{20}_{489}$ +20° (*c* 0.13, $CHCl_3$))
	cyclo- [e.g., *cyclo*-hexasulfur (*c*-S_6); *cyclo*-(alanylglycyl)]
C	Celsius
	coulomb
	cytidine
	ring (italicize in names of steroids)
C	chair (in ring conformation)
	clockwise
ca.	approximately (circa; about)
CAD	computer-assisted design
cal	calorie
cal$_{IT}$	International Table calorie
calcd	calculated
CAM	computer-assisted makeup
cAMP	adenosine cyclic 3′,5′-phosphate
	adenosine 3′,5′-cyclic phosphate
	adenosine 3′,5′-phosphate
CAN	ceric ammonium nitrate

CARS	coherent anti-Stokes Raman spectroscopy
CAT	computer-averaged transients
cB	conjugate (counter) base (also CB, cb)
Cbz	carbobenzoxy, carbobenzyloxy, (benzyloxy)carbonyl, benzyloxycarbonyl
cc	cubic centimeter (*use* cm^3)
cd	candela
	current density
CD	circular dichroism
CDH	ceramide dihexosides ($Cer(Hex)_2$)
CDP	cytidine 5′-diphosphate
CE	Cotton effect
cf.	compare
cfm	cubic feet per minute
cfse	crystal field stabilization energy (also CFSE)
cfu	colony-forming units (bacterial inocula)
cgs	centimeter–gram–second system
cgsu	centimeter–gram–second unit(s)
ChE	cholinesterase
CHF	coupled Hartree–Fock
Ci	curie
CI	chemical ionization
	configuration interaction
CIDEP	chemically induced dynamic electron polarization
CIDNP	chemically induced dynamic nuclear polarization
CL	chemiluminescence
	cathodoluminescence
clc	capital and lower case (alphabet letters)
c/m^2	candles per square meter
CM	carboxymethyl (as in CM-cellulose)
cmc	critical micelle concentration (*do not use* CMC)
CMH	ceramide monohexosides [Cer(Hex)]
CMO	canonical molecular orbital
CMP	cytidine 5′-monophosphate
	cytidine 5′-phosphate
cmr	*use* ^{13}C NMR
CN	coordination number
CNDO	complete neglect of differential overlap (CNDO/2, etc.)
CNS	central nervous system
co	copoly (as in A-*co*-B)
CoA	coenzyme A
COD	chemical oxygen demand
	cyclooctadiene (also cod)
coeff	coefficient
colog	cologarithm
compd	compound
con	conrotatory (may be italic)

concd	concentrated
concn	concentration
const	constant
cor	corrected
cos	cosine
cosh	hyperbolic cosine
cot	cotangent
coth	hyperbolic cotangent
count/s	counts per second (s^{-1} is preferred)
C_p	heat capacity at constant pressure
cp	candlepower
cP	centipoise
Cp	cyclopentadienyl (also cp)
CP	chemically pure
	cross polarization, cross-polarization
cpd	contact potential difference
CPE	controlled-potential electrolysis
CPK	Corey-Pauling-Koltun (models)
CPL	circular polarization of luminescence
cpm	counts per minute
CP/MAS	cross-polarization/magic angle spinning (other variations, e.g., using hyphens, are permitted—CP-MAS, CPMAS, CP MAS)
cps	cycles per second (*use* Hz)
	counts per second (*use* counts/s *or* s^{-1})
CPU	central processing unit
crit	critical
cRNA	complementary RNA
CRT	cathode ray tube
CRU	constitutional repeating unit
cryst	crystalline
csc	capital and small capital (alphabet letters)
	cosecant
csch	hyperbolic cosecant
CT	charge transfer
CTH	ceramide trihexosides [Cer(Hex)$_3$]
CTP	cytidine 5′-triphosphate
cub	cubic (also c)
C_v	heat capacity at constant volume
CV	cyclic voltammetry
CW	constant width
	continuous wave (as in CW ESR)
cwt	hundredweight
Cy	cyclohexyl (also cHx, c-Hx, and others)
cyclam	1,4,8,11-tetraazacyclotetradecane
Cyd	cytidine
Cys	cysteinyl, cysteine (C)
cytRNA	cytoplasmic RNA

d	day (*do not use; spell out* day)
	deci (10^{-1})
	deoxy
	deuteron
	differential (mathematical)
	diffuse
	doublet (spectral)
d.	diameter, with i. and o. (inside and outside)
d	density
	dextrorotatory
	distance
	spacing (X-ray)
D	configurational
δ	scale (NMR), dimensionless
D	debye
	ring (italic in names of steroids)
D	diffusion coefficient ($cm^2 s^{-1}$)
	symmetry group [e.g., D_3; also used in names, such as (+)-D_3-trishomocubane]
3-D	three-dimensional (also 3D)
da	deka (10)
daf	dry ash free
dAMP	2′-deoxyadenosine 5′-phosphate, etc. (the etc. stands for replacing the A with C, G, U, etc.)
dansyl	8-(dimethylamino)-1-naphthalenesulfonate
dB	decibel
dc	direct current
DD-8	diagonal dodecahedron
DDT	1,1,1-trichloro-2,2-bis(*p*-chlorophenyl)ethane
DEAE	(diethylamino)ethyl (as in DEAE-cellulose)
dec	decomposition
decomp	decompose
DEFT	driven equilibrium Fourier transform
deg	degree (°B, degrees Baume; °C, °F, but K)
DEG	diethylene glycol
DEHP	bis(2-ethylhexyl) phthalate (*change abbreviation to* BEHP)
DES	diethylstilbestrol
det	determinant
df	degrees of freedom
dil	dilute
dis	disrotatory (may be italic)
distd	distilled
DMBA	9,10-dimethylbenz[*a*]anthracene
DME	dropping mercury electrode
DMF	dimethylformamide
dmr	*use* ^{2}H NMR
DMSO	dimethyl sulfoxide (*change to* Me_2SO or $(CH_3)_2SO$)
DNA	deoxyribonucleic acid

DNase	deoxyribonuclease
DNMR	dynamic nuclear magnetic resonance
Dnp	dinitrophenyl (N_2ph is a recommended alternative)
DNP	deoxynucleoprotein
	dinitrophenylhydrazone
Dns	dansyl (also DNS)
Dopa	dihydroxyphenylalanine (also DOPA)
dp	degree of polymerizaton
dpm	disintegrations per minute
DPN	diphosphopyridine nucleotide (NAD)
DPNH	reduced DPN
DPPH	diphenylpicrylhydrazyl
dps	disintegrations per second
Dq	crystal field splittings (also *Ds*, *Dt*)
DSC	differential scanning calorimetry
Dt	crystal field splittings
DTA	differential thermal analysis
DTE	dithioerythritol
DTT	dithiothreitol
DTC	differential thermal calorimetry
	depolarization thermocurrent
dyn	dyne
e	electron
	equatorial (in names, *use as* 2(e)-methyl)
e_{aq}^-	hydrated electron [also $e^-(aq)$]
e_s^-	solvated electron [also $e^-(s)$]
e	base of natural logarithm
E	electromotive force (also E_{MF})
	entgegen (configuration)
	envelope (conformation)
	specific extinction coefficient (as $E_{280nm}^{1\%,\ 1\ cm}$)
E2	elimination, second order
ϵ	molar absorptivity
η	viscosity
ea_0	electronic charge in esu x Bohr radius or au for dipole moment
$e\ Å^{-3}$	electrons per cubic angstrom
EC	Enzyme Commission
ECE	electrochemical, chemical, electrochemical (mechanisms)
ECG	electrocardiogram (also EKG)
ecl	electrochemical luminescence
ed.	edition
Ed., Eds.	editor, editors
ED	effective dose
ED_{50}	dose that is effective in 50% of test subjects (also ED50)
edda	ethylenediaminediacetate
EDTA	ethylenediaminetetraacetic acid, -tetracetate, -tetraacetato (also edta as ligand)
ee	enantiomeric excess

EEG	electroencephalogram
EFG	electric field gradient
e.g.	for example
EGA	evolved gas analysis
EGD	evolved gas detection
EGR	exhaust gas recirculation
EH	extended Hückel
EI	electron impact
	electron ionization
E_k	kinetic energy (also KE)
EKG	electrocardiogram (also ECG)
EL	electroluminescence
EM	electron microscopy
emf	electromotive force
emu	electromagnetic unit
en	ethylenediamine
ENDOR	external nuclear double resonance
ent	reversal of stereo centers
E_p	potential energy (also PE)
EPA	ether–isopentane–ethanol (solvent system)
epi	inversion of normal configuration (italic with a number, as in 15-*epi*-prostaglandin A; generally roman without)
EPR	electron paramagnetic resonance
eq	equation
equiv	equivalent
equiv wt	equivalent weight
erf	error function
erfc	erf complement
$erfc^{-1}$	inverse erfc
ESCA	electron spectroscopy for chemical analysis (also XPS)
esd	estimated standard deviation
ESE	electron spin echo
ESEEM	electron spin echo envelope modulation
ESP	elimination of solvation procedure
	extrasensory perception
ESR	electron spin resonance
esu	electrostatic unit
Et	ethyl
et al.	and others
etc.	and so forth
eu	entropy unit
EU	enzyme unit
eV	electronvolt
EXAFS	X-ray absorption fine structure, extended
exp	exponential
expt(l)	experiment(al)
f	femto (10^{-15})
	fine (spectral)
	function, as f(x) [also $f(x)$]

f	focal length
	furanose form
F	farad
	fermi
	formal (*use judiciously;* M is preferred)
F	Faraday constant (also $\mathscr{F}$)
	free energy
FAAS	flame atomic absorption spectroscopy
FABMS	fast atom bombardment mass spectrometry
fac	facial
FAD	flavin adenine dinucleotide
FAES	flame atomic emission spectroscopy
FAFS	flame atomic fluorescence spectroscopy
FAS	flame absorption spectroscopy
FCC	fluid catalytic cracking
fcc	face-centered cubic (also fccub)
Fd	ferredoxin
FEM	flame emission spectroscopy
ff	and following (as in p 457 ff)
FFEM	freeze-fracture electron microscopy
FFS	flame fluorescence spectroscopy
FFT	fast Fourier transform
FHT	Fisher–Hirschfelder–Taylor space-filling models
FI	field ionization
fid	free induction decay (in Fourier transform work)
FIK	field ionization kinetics
FIR	far infrared (*use* far-IR or spell out)
fL	footlambert
FM	frequency modulation
FMN	flavin mononucleotide
FMO	frontier molecular orbital
FOPPA	first-order polarization propagator approach
fp	freezing point
FPC	fixed partial charge
FPT	finite perturbation theory
Fr	franklin
Fr	Froude number
Fru	fructose
FSGO	floating spherical Gaussian orbital
FSH	follicle-stimulating hormone
ft	foot
FT	Fourier transform
FT IR	Fourier transform infrared
	FT/IR, FT-IR, FTIR all acceptable
ft-c	foot-candle
ft lbf	foot pound-force
FTS	Fourier transform spectroscopy (*do not use*)
fw	formula weight

fwhm	full width at half-maximum
g	gas (as in H_2O_g)
	gram
(g)	gas [as in $H_2O(g)$]
g	gravitation constant
γ	microgram (*use* μg)
	wavelength
Γ	ionic strength
G	gauss
	generally labeled
	giga (10^9)
	guanosine
G	free energy (Gibbs)
Ga	Galileo no.
gal	gallon
Gal	galactose
g-atom	gram atom (*use* mol)
GC	gas chromatography
GDC	gas displacement chromatography
GDP	guanosine diphosphate
GFC	gas frontal chromatography
gfw	gram formula weight (*do not abbreviate*)
Gi	gilbert
GIAO	gauge-invariant atomic orbital
Glc	glucose
GLC	gas–liquid chromatography
Gln	glutaminyl, glutamine (Q)
Glu	glutamyl, glutamic acid (E)
Glx	Gln or Glu
Gly	glycyl, glycine (in coordination names, gly; G)
GMP	guanosine monophosphate
GPC	gel permeation chromatography
gr	grain (unit of weight)
GSC	gas–solid chromatography
GSH	reduced glutathione
GSL	glycosphingolipid
GSSG	oxidized glutathione
GTP	guanosine 5′-triphosphate
Guo	guanosine
h	hecto (10^2)
	hour
h	h index (*hkl*)
	Planck constant
$\hbar$	Planck constant divided by 2π
H	henry
H	enthalpy
	half-chair (conformational)
	Hamiltonian (also $\mathscr{H}$)

ha	hectare
Hb	hemoglobin
Hbg	biguanide
hcp	hexagonal close-packed
HCP	hexachlorophene
Hedta	ethylenediaminetetraacetate(3-) (ato as ligand in full name)
H_2edta	ethylenediaminetetraacetate(2-) (ato as ligand in full name)
H_3edta	ethylenediaminetetraacetate(1-) (ato as ligand in full name)
H_4edta	ethylenediaminetetraacetic acid
HEEDTA	N-(2-hydroxyethyl)ethylenediaminetriacetate
Hepes	N-(2-hydroxyethyl)piperazine-N'-2-ethanesulfonic acid (also HEPES, hepes)
HEPPS	N-(2-hydroxyethyl)piperazine-N'-3-propanesulfonic acid (also Hepps, hepps)
hex	hexagonal (crystal structure)
hfs	hyperfine splitting
hfsc	hyperfine splitting constant
His	histidyl, histidine (H)
HMDS	hexamethyldisilane
	hexamethyldisiloxane
HMO	Hückel molecular orbital
HMPA	hexamethylphosphoramide, hexamethylphosphoric triamide (also HMPT)
hn	heterogeneous nuclear (as in hnRNA)
HOMO	highest occupied molecular orbital
H_2ox	oxalic acid
hp	horsepower
HPLC	high-pressure liquid chromatography
	high-performance liquid chromatography
Hyl	hydroxylysyl, hydroxylysine
Hyp	hydroxyprolyl, hydroxyproline
Hz	hertz (s^{-1})
i	iso (as in i-Pr; never i-propyl)
I	electric current
	ionic strength
	moment of inertia
IA	international angstrom
ibid.	in the same place (in the reference cited)
ic	intracerebrally
ICR	ion cyclotron resonance
ics	internal chemical shift
ICSH	interstitial cell-stimulating hormone
ICT	International Critical Tables
i.d.	inside diameter
i_d	diffusion current
ID	infective dose
ID_{50}	dose that is infective in 50% of test subjects (also ID50)
IDP	inosine 5'-diphosphate

i.e.	that is
IE	ionization energy
IEC	ion-exchange chromatography
IKES	ion kinetic energy spectroscopy
Ile	isoleucyl, isoleucine (I)
ILS	increased life span (*do not generally use*)
im	intramuscularly
IMP	inosine 5'-phosphate
in.	inch
INDO	intermediate neglect of differential overlap
	internucleus (nucleus–nucleus)
INDOR	internal nuclear double resonance
	internucleus (nucleus–nucleus) double resonance
Ino	inosine
INO	iterative natural orbital
insol	insoluble
ip	intraperitoneally
IP	ionization potential
ips	iron pipe size
ipso	position of substitution
IR	infrared
IRDO	intermediate retention of differential overlap
IRP	internal reflection photolysis
IRS	internal reflection spectroscopy
isc	intersystem crossing
ISCA	ionization spectroscopy for chemical analysis
iso	inversion of normal chirality (not as in isopropyl, but in uses such as 8-*iso*-prostaglandin E_1; generally with a number)
ITP	inosine 5'-triphosphate
IU	international unit
iv	intravenously
J	joule
k	kilo (10^3)
k	Boltzmann constant
	k index
	rate constant
K	kayser (*use* cm^{-1})
	kelvin (formerly °K)
	1000 (as in 60K protein)
	1024 (computer terminology; as in 8K or 16K disk drives)
K	equilibrium constant
K_m	Michaelis constant (also K_M)
kat	katal
kb	kilobar (*use* kbar)
	kilobase
kbar	kilobar
kbp	kilobase pair

KE	kinetic energy (also E_k)
kX	crystallographic unit, ca. 1 Å
l	liquid (as in HCl_l)
	liter (*change to* L)
(l)	liquid [as in $NH_3(l)$]
λ	microliter (*use* μL)
l	*l* index
	levorotatory
L	configuration (rare)
L	configuration
L	lambert
	ligand
	liter
Lac	lactose
lat	latitude
lb	pound
lc	lower case (alphabet letters)
LC	liquid chromatography
LCAO	linear combination of atomic orbitals
LCICD	liquid crystal induced circular dichroism
LCVAO	linear combination of virtual atomic orbitals
LD	lethal dose
LD_{50}	dose that is lethal to 50% of test subjects (also LD50)
LE	locally excited (cf. CT)
LEED	low-energy electron diffraction
LEMF	local effective mole fraction
Leu	leucyl, leucine (L)
LFER	linear free energy relationship
LH	luteinizing hormone
lim	limit
LIS	lanthanide-induced shift
lit.	literature
LJ	Lennard-Jones
LLC	liquid–liquid chromatography
lm	lumen
ln	natural logarithm
LNDO	local neglect of differential overlap
log	logarithm to the base 10
long.	longitude
Lp	Lorentz–polarization
LSC	liquid–solid chromatography
LSD	lysergic acid diethylamide
LSR	lanthanide shift reagent
LUMO	lowest unoccupied molecular orbital
lut	lutidine
Lys	lysyl, lysine (K)
lx	lux

m	medium (spectra)
	meter
	mile (in mph; but usually mi)
	milli (10^{-3})
	month (spell out)
	multiplet (spectra)
μ	micro (10^{-6})
	micron (*use* μm)
μ_B	Bohr magneton
m	meta
	molal
M	mega (10^6)
	mesomeric
	metal
	mol dm^{-3}, mol L^{-1} (molar)
M	minus (left-handed helix)
[M]	molecular rotation
Mal	maltose
Man	mannose
MAO	monoamine oxidase
MASS	magic angle sample spinning
max	maximum
mb	myoglobin
MCD	magnetic circular dichroism
MD	molar rotation
m/e	mass-to-charge ratio (also *m/z*)
Me	methyl
MED	mean effective dose
MEM	minimum Eagle's essential medium
mer	meridional
Mes	mesylate
MES	2-morpholinoethanesulfonic acid, -sulfonate (also Mes)
Met	methionyl, methionine (M)
MeV	million electron volts
mf	mole fraction (*discouraged*)
MF-FSGO	molecular fragment
mi	mile
min	minimum
	minute
MINDO	modified intermediate neglect of differential overlap
MIR	medium infrared (*avoid*)
MIRS	multiple internal reflection spectroscopy
mmHg	measure of pressure, related to mercury
mmp	mixture melting point
mmu	millimass unit
M_n	number-average molecular weight
mo	month

MO	molecular orbital
mol	mole
mol wt	molecular weight (*use* M_r or MW when appropriate)
mon	monoclinic (crystal structure)
mp	melting point
MPI	multiphoton ionization
MPV	Meerwein–Ponndorf–Verley
M_r	relative molecular mass, "molecular weight"
MR	molecular refraction
mRNA	messenger RNA
MS	mass spectroscopy
	mass spectrum
MSH	melanocyte-stimulating hormone, melanotropin
MTD	mean therapeutic dose
mtRNA	mitochondrial RNA
mu	mass units (*generally spell out*)
MVA	mevalonic acid
MVS	multiple-variable storage (an operating system)
MW	molecular weight (also mol wt, M_r)
M_w	weight-average molecular weight
Mx	maxwell
M_z	z-average molecular weight
m/z	mass-to-charge ratio (also m/e)
n	nano (10^{-9})
	neutron
n	normal (as in *n*-butyl, *n*-Bu)
	refractive index (n^{20}_D, at 20 °C, Na D line)
ν	frequency (wavenumber)
N	isotopic labeling (when the labeling method results in most of the label residing in the designated site, but no information is available on the extent of labeling of other sites)
	newton
	normal (concentration)
	unspecified nucleoside
[Na]ATPase	sodium ion activated ATPase (likewise, [Na,K]ATPase)
NAD	nicotinamide adenine dinucleotide (also DPN)
NADH	reduced NAD
NADP	NAD phosphate (also TPN)
NADPH	reduced NADP
N.B.	nota bene (note well)
NBS	*N*-bromosuccinimide (also SucNBr)
NDDO	neglect of diatomic differential overlap
NEMO	nonempirical molecular orbital
neut equiv	neutralization equivalent
NHE	normal hydrogen electrode
NIR	near infrared (*do not use*)
Nle	norleucyl, norleucine

nm	nanometer
	nuclear magneton (*use* μ_N)
NMN	nicotinamide mononucleotide
NMR	nuclear magnetic resonance
no.	number
NO	natural orbital (as in CNDO/2-NO)
NOCOR	neglect of core orbitals
NOE	nuclear Overhauser effect
Np	neper
NPR	net protein retention
NQI	nuclear quadrupole interaction
NQR	nuclear quadrupole resonance
nRNA	nuclear RNA
NRTL	nonrandom two-liquid
NTP	normal temperature and pressure
	unspecified nucleoside 5′-triphosphate
o	ortho
Ω	ohm
OAc	acetate (also AcO^-)
obsd	observed
OC-6	octahedral, CN 6
o.d.	outside diameter
OD	optical density
ODMR	optically detected magnetic resonance
ODU	optical density units
Oe	oersted
OFDR	off-frequency decoupling resonance
ORD	optical rotatory dispersion
o-rh	orthorhombic (crystal structure)
Orn	ornithyl, ornithine
osM	osmolar (also osm)
ox	oxalate
	oxidized or oxidation (in subscripts and superscripts)
oxidn	oxidation
oz	ounce
p	page
	pico (10^{-12})
	proton
	negative logarithm of (as in pH, pa_H, pcH)
p	para
	pyranose form
P	poise
P	plus (right-handed helix)
P450	as in cytochrome P450 (also P-450, P_{450})
π	pros (near) in NMR measurements (as in N^π of histidine)
P_i	inorganic phosphate
Pa	pascal
pa_H	negative logarithm of hydrogen ion activity

PAGE	polyacrylamide gel electrophoresis
PBS	phosphate-buffered saline (*use* NaCl-*P* buffer if necessary; specify pH and concentration)
PC	paper chromatography
PCB	polychlorobiphenyl
	polychlorinated biphenyl
p*c*H	measure of hydrogen concentration
PCILO	perturbed configuration interaction with localized orbitals
PCP	pentachlorophenol
pd	potential difference
PE	potential energy (also E_p)
%	percent (parts per hundred)
‰	per thousand (parts per thousand)
PES	photoelectron spectroscopy
PET	positron emission tomography
PFU	plaque-forming units
PG	prostaglandin (as PGA, PGB, etc.; *do not generally use*)
Ph	phenyl
pH	hydrogen ion concentration
Phe	phenylalanyl, phenylalanine (F)
phen	1,10-phenanthroline, *o*-phenanthroline
pHs	acceptable plural for pH (also pH's, pH values)
pK	negative logarithm of equilibrium constant
pK_a	pK for association
pKs	acceptable plural for pK (also pK's, pK values)
PL	photoluminescence
p.m.	post meridiem
PMO	perturbational molecular orbital
PMR	polymerization of monomeric reactants
	proton magnetic resonance; change to ^{1}H NMR
PNDO	partial neglect of differential overlap
po	per os (orally)
POPOP	1,4-bis(5-phenyl-2-oxazolyl)benzene
pp	pages
ppb	parts per billion
PP$_i$	inorganic pyrophosphate, -phosphoric acid, etc.
ppm	parts per million
PPO	2,5-diphenyloxazole
PPP	Pariser–Parr–Pople
PPS	photophoretic spectroscopy
ppt	precipitate
Pr	propyl
ψrd	pseudouridine
PRDDO	partial retention of diatomic differential overlap
PRF	Petroleum Research Fund
PRFT	partially relaxed Fourier transform
Pro	prolyl, proline (P)
pro-R	stereochemical descriptor

pro-S	stereochemical descriptor
PRT	platinum resistance thermometer
Ps	positronium
psi	pounds per square inch
psia	pounds per square inch absolute
psig	pounds per square inch gauge
pt	pint
PTC	phase-transfer catalysis
PTFE	poly(tetrafluoroethylene)
PTH	parathyroid hormone
	phenylthiohydantoin
py	pyridine (coordination names)
PY	pyramidal
pyr	pyridine
	pyrazine (in coordination names; not good, because Pyr = pyr = pyridine for most organic chemists)
pyrr	pyrrolidine (coordination names)
pz	pyrazine
q	quartet (in spectra)
QCPE	Quantum Chemistry Program Exchange
qt	quart
ρ	density (also d, other variations)
r	reference (as in r-2,c-3,t-4-...)
R	roentgen
R	rectus (configurational)
	resistance
rac	racemic
rad	radian
	unit of radiation (also rd)
rd	rad
re	stereochemical descriptor (as in the *re* face)
recryst	recrystallized
red	reduced or reduction (in subscripts and superscripts)
redn	reduction
ref	reference
rel	relative
rem	roentgen equivalent man
REM	rapid eye movement
rep	roentgen equivalent physical
rf	radio frequency
R_f	retardation factor (ratio of distance traveled by the center of a zone to the distance simultaneously traveled by the mobile phase)
Rib	ribose
rms	root mean square (spell out unless used frequently)
RNA	ribonucleic acid (also tRNA, mRNA, etc.)
RNase	ribonuclease
ROA	Raman optical activity

RPLC	reversed-phase liquid chromatography
rpm	revolutions per minute
RQ	respiratory quotient
RRDE	rotating ring-disk electrode
RRKM	Rice–Ramsperger–Kassel–Marcus theory
rRNA	ribosomal RNA
Ry	rydberg
s	second
	single bond [as in s-cis (italic in compound names)]
	singlet (NMR)
	solid (as in $NaCl_s$)
	strong (IR)
(s)	solid [as in $I_2(s)$]
s	sedimentation coefficient
	standard deviation (analytical)
	symmetrical
σ	standard deviation
S	siemens (conductance)
	svedberg
S_{ex}	exciplex substitution
S	entropy
	sinister (configurational)
	skew (conformation)
$s^0_{20,w}$	sedimentation coefficient measured at 20 °C in water and extrapolated to 0 °C
SA-8	square antiprism, CN 8
Sar	sarcosyl, sarcosine (*N*-methylglycine); sar as ligand
SAR	structure–activity relationship
sc	subcutaneously
sc	synclinal
SCE	standard calomel electrode
SCF	self-consistent field
scfh	standard ft^3/h
SD	standard deviation
SDS	sodium decyl (or dodecyl) sulfate
SE	standard error
$S_E 2$	electrophilic substitution, second order
sec	secant
sec	secondary (as in *sec*-butyl, *sec*-Bu)
SECS	simulation and evaluation of chemical synthesis
SEM	standard error of the mean
Ser	seryl, serine (S)
sh	sharp (spectral)
	shoulder (spectral)
SH	Sherwood number (italic in math)
SHC	shape and Hamiltonian consistent
SHE	standard hydrogen electrode

si	stereochemical descriptor (as in the *si* face)
SI	International System of Units
	secondary ion (as in SIMS)
SIMS	secondary ion mass spectrometry
sin	sine
sinc	sin x/x
sinh	hyperbolic sine
SLR	spin–lattice relaxation
sn	stereospecific numbering
S_N1	nucleophilic substitution, first order
S_N2	nucleophilic substitution, second order
S_Ni	nucleophilic substitution, internal
SNO	semiempirical natural orbital
sol	solid
soln	solution
sp	specific
sp	synperiplanar
SP	square planar
sp gr	specific gravity
sp ht	specific heat
SPR	stroboscopic pulse radiolysis
sp vol	specific volume
SPY	square pyramidal
SP-4	square planar, CN 4
SP-5	square pyramid, CN 5
sq	square
sr	steradian
SSC	standard saline citrate (NaCl–citrate)
St	stokes
std	standard
STEM	scanning transmission electron microscope
STO	Slater-type orbitals
STO-3G	Slater-type orbitals, three Gaussian
STP	standard temperature and pressure
Suc	sucrose
SucNBr	*N*-bromosuccinimide (also NBS)
Sv	Sievert
SVL	single vibrational level
swg	standard wire gauge
sym	symmetrical
t	triplet (spectra)
	triton
t	temperature, °C
	tertiary (as in *t*-Bu; but *tert*-butyl)
	trans (stereochemical descriptor)
T	tautomeric
	tera (10^{12})
	tesla (Wb/m^2)

T	temperature, K
	twist (conformation)
T-4	tetrahedral, CN 4
τ	tele (far) in NMR measurements (as in N^{τ} of histidine)
θ	ORD measurement, deg $cm^2/dmol$
tan	tangent
tanh	hyperbolic tangent
TBP	tri-*n*-butyl phosphate
	trigonal bipyramidal
T/C	treated vs. cured
TCA	tricarboxylic acid cycle (citric acid cycle, Krebs cycle)
	trichloroacetic acid (also Cl_3CCOOH or Cl_3AcOH)
TDS	total dissolved solids
TEA	transversely excited atmospheric
TEAE	triethylaminoethyl (as in TEAE-cellulose)
temp	temperature
tert	tertiary (as in *tert*-butyl; but *t*-Bu)
tetr	tetragonal (crystal structure)
TFA	trifluoroacetyl (also F_3Ac)
TGA	thermogravimetric analysis
TH	trihydro; tetrahydro (*but do not use; preferred usage is* H_3, H_4)
Tham	Tris
THC	tetrahydrocannabinol (also H_4Can)
theor	theoretical
THF	tetrahydrofuran (also H_4furan)
Thr	threonyl, threonine
TIP	temperature-independent paramagnetism
TL	triboluminescence
TLC	thin-layer chromatography
TM	trimethyl, tetramethyl (preferred: Me_3, Me_4)
TMS	tetramethylsilane (*use* Me_4Si *or* $(CH_3)_4Si$)
	trimethylsilyl (*use* Me_3Si *or* $(CH_3)_3Si$)
TMV	tobacco mosaic virus
TnL	tunnel luminescence
TOD	total oxygen demand
tol	toluene
TPN	triphosphopyridine nucleotide (NADP)
TPNH	reduced TPN
TP-6	trigonal prism, CN 6
t_R	retention time
Tr	trace
tric	triclinic (crystal structure)
triflate	trifluoromethanesulfonate
trig	trigonal (crystal structure)
Tris	tris(hydroxymethyl)aminomethane
tRNA	transfer RNA
Trp	tryptophyl, tryptophan (W)
TSC	thermal stimulated current

TSH	thyroid-stimulating hormone
Tyr	tyrosyl, tyrosine (Y)
u	unified atomic mass unit
U	uniformly labeled
UDP	uridine 5′-diphosphate
uhf	ultrahigh frequency
UHF	unrestricted Hartree–Fock
UMP	uridine 5′-phosphate
uncor	uncorrected
uns	unsymmetrical
UPS	ultraviolet photoelectron spectroscopy (also UV PES)
ur	urea (abbreviate as ligand only)
Urd	uridine
USP	United States Pharmacopeia
UTP	uridine 5′-triphosphate
UV	ultraviolet
UV PES	ultraviolet photoelectron spectroscopy (also UPS)
UV–vis	ultraviolet–visible
v	vicinal (also *vic*)
V	volt
Val	valyl, valine (V)
VASS	variable angle sample spinning
VB	valence bond
VCD	vibrational circular dichroism
VESCF	variable electronegativity self-consistent field
vic	vicinal (also *v*)
viz.	namely
VLE	vapor–liquid equilibrium
VOA	vibrational optical activity
vol	volume
vol %	volume percent
vp	vapor pressure
vpc	vapor phase chromatography (*do not use*)
VPC	vapor pressure chromatography
VPO	vapor pressure osmometry
vs.	versus
vs	very strong (spectral)
VSIP	valence state ionization potential
VUV	vacuum ultraviolet (*avoid*)
v/v	volume-to-volume ratio
VVk	van Vleck
vw	very weak (spectral)
w	weak (spectral)
W	watt
Wb	weber
wrt	with respect to (*do not use*)
wt	weight
wt %	weight percent

w/w	weight-to-weight ratio
X	xanthosine (*use* N for unknown nucleoside)
Xao	xanthosine
XMP	xanthosine 5'-phosphate
XPS	X-ray photoelectron spectroscopy (also ESCA)
XRD	X-ray diffraction
Xyl	xylose
yr	year (acceptable in limited circumstances; "a" is official abbreviation; generally spell out)
Z	zusammen (configurational)
zfs	zero-field splitting
zfsc	zero-field splitting constant

Mathematical Symbols

By mathematical symbols, we mean single letters used to designate unknown quantities, constants, and variables. See also the discussion under "Abbreviations and Acronyms" in which the differences between abbreviations and mathematical symbols are described.

- Mathematical symbols are set in italic type, except vectors and tensors, which are generally set in boldface Roman type. In typewritten copy, often an arrow is drawn above the vector symbol. This arrow is deleted in typesetting, and the symbol is marked for boldface. An arrow over an italic letter is acceptable notation for a vector, but it should be used only when the boldface representation has been assigned to a tensor.

- Numerals, operators, and punctuation are not italic, nor are trigonometric functions or abbreviations.

cos	cosine	log	logarithm
cosh	hyperbolic cosine	Log	principal log
cot	cotangent	max	maximum
coth	hyperbolic cotangent	min	minimum
det	determinant	mod	modulus
dim	dimension	sin	sine
exp	exponential	sinh	hyperbolic sine
lim	limit	tan	tangent
ln	natural logarithm	tanh	hyperbolic tangent

- In typewritten manuscript, use an underline for italic and a wavy line for bold. Never underline Greek letters and do not use underlines or wavy lines unless your usage coincides with that of your publisher.

- Items set in Roman type are preceded and followed by a space.

log 2	*not*	log2
$-\log x$	*not*	$-\log x$
4 sin θ	*not*	4sinθ

  Exception: exp($-x$)

  More complicated arguments should also be placed within brackets.

  $$\sin\left[2p(x-y)/n\right] \qquad \log\left[-V(r)/kT\right]$$

- Leave a space before and after mathematical operators that have numbers on both sides.

  $$20 \pm 2\% \qquad 3.24 \pm 0.01 \qquad 4 \times 5 \text{ cm}$$

- When plus or minus signs are used with one number that is not part of an equation, do not leave a space between the sign and the number.

  -12 °C   25 g ($\pm1\%$)   The level can vary from -15 to $+25$ m.

- Mathematical expressions can be used as part of a sentence when the subject, verb, and object are all part of the mathematical expression.

  When $V = 12$, then eq 15 is valid.

  (V is the subject, $=$ is the verb, and 12 is the object.)

  However, do not use the equals sign instead of the word "equals" in narrative text. The same applies to the greater than and less than symbols and plus and minus signs.

$PV = nRT$ where P is pressure	*not*	where $P =$ pressure
when the temperature is 50 °C	*not*	when the temperature $=$ 50 °C
when A and B are reacted	*not*	when A $+$ B are reacted

- Form the plurals of mathematical symbols by adding an apostrophe and "s" if you cannot use a word like "values" and "levels".

at r values greater than	*is better than*	at r's greater than

- In equations, leave a space before and after mathematical operators ($+$, $-$, $\times$, $=$) except when they are superscripts or subscripts.

- Two different characters in typeset type may be represented by only one character on many typewriters and printers. Or, they may be so similar as to be indistinguishable: for example, the letter *l* and the number *1*, the letter *O* and *0*, and prime and apostrophe. These characters should be spelled out in the margin each time they appear, or the typesetter will not know which one is meant. For other similar letters, such as *v* and the Greek ν, *k* and the Greek κ, *u* and the Greek μ, and *n* and the Greek η, the Greek letter need be specified only the first time it appears in text.

- Equations displayed on separate lines should be numbered with Arabic numbers in parentheses placed at the right margin. If an equation is very short and will not be referred to again, it should be run into text and not numbered. Use parentheses in accordance with the rules of mathematics. If the solidus (/) is used in division and if there is any doubt where the numerator starts or where the denominator ends, use parentheses.

    $(a/b)/c$ *or* $a/(b/c)$ *never* $a/b/c$

    $$\frac{x+y}{2} = z$$ *would be better as* $(x+y)/2 = z$

    $$\frac{x-y}{z} + 2a$$ *would be better as* $[(x-y)/z] + 2a$

- Do not use any punctuation after equations displayed on separate lines.

- Enclose parentheses within square brackets, and square brackets within braces.

    $$z = 10\{(x+y) + c[2(a-b) + 5]\}$$

- Do not use square brackets, parentheses, or braces around the symbol for a quantity to make it represent any other quantity.

    Incorrect
    where P is pressure and (P) is pressure at equilibrium

    Correct
    where P is pressure and P_e is pressure at equilibrium

- Use square brackets to denote coordination entities.

    $[Cr(C_6H_6)_2]$ $K[PtCl_3(C_2H_4)]$

- Use square brackets enclosing a formula of a chemical species to indicate its concentration in reactions and equations, but not in narrative text.

  Correct
  $[Mg] = 3 \times 10^{-2}$ M

  Incorrect
  The [Mg] was found to be greater in the unwashed samples.

- In crystallography, use parentheses to enclose Miller indices, do not enclose Laue indices, enclose indices of a form in braces, and enclose indices of a zone axis or line in square brackets. A symbol such as 123 or *hkl* is understood to be a reflection, (123) or (*hkl*) is a plane or set of planes, [123] or [*uvw*] is a direction, {*hkl*} is a form, and ⟨*uvw*⟩ represents all crystallographically equivalent directions of the type [*uvw*].

- Some usages and symbols for mathematical operations and constants are the following:

$\approx$	approximately equal to
$\sim$	asymptotically equal to
$\propto$	proportional to
$\rightarrow$	approaches (tends to)
$\|a\|$	absolute magnitude of a
$\equiv$	identically equal to
$a^{1/2}, \sqrt{a}, \sqrt{a}$	square root of a
$a^n, \sqrt[n]{a}, \sqrt[n]{a}$	nth root of a
$\overline{a}, \langle a \rangle$	mean value of a
$\log x$	logarithm to the base 10 of x
$\log_a x$	logarithm to the base a of x
$\ln x$	natural logarithm of x
$\exp x, e^x$	exponential of x
Δx	finite increment of x
∂x	infinitesimal increment of x
dx, dx	total differential of x
$f(x)$	function of x
$\int y \, dx$	integral of y with respect to x
$\int_a^b y \, dx$	integral of y from $x = a$ to $x = b$
A	vector of magnitude A
A·B	scalar product of **A** and **B**
A $\times$ **B**, **AB**	vector product of **A** and **B**
A	matrix A
∞	infinity
Σ	summation
Π	product

The Greek Alphabet

Name	Capital	Lower Case	Name	Capital	Lower Case
Alpha	A	α	Nu	N	ν
Beta	B	β	Xi	Ξ	ξ
Gamma	Γ	γ	Omicron	O	o
Delta	Δ	$\delta\ \partial^a$	Pi	Π	π
Epsilon	E	ϵ	Rho	P	ρ
Zeta	Z	ζ	Sigma	Σ	σ
Eta	H	η	Tau	T	τ
Theta	Θ	$\theta\ \vartheta^a$	Upsilon	Υ	υ
Iota	I	ι	Phi	Φ	$\phi\ \varphi^a$
Kappa	K	κ	Chi	X	χ
Lambda	Λ	λ	Psi	Ψ	ψ
Mu	M	μ	Omega	Ω	ω

^{a} Old style; still used in mathematics.

Statistics

Certain statistical symbols are standard:

$\overline{x}$	arithmetic mean	CV	coefficient of variation
r	correlation coefficient	Σ	summation
R	regression coefficient	SE	standard error
n, N	total number of individuals or variates	SEM	standard error of the mean
f	frequency	t	Student distribution (the Student t test)
p, P	probability	F	variance ratio
σ, SD	standard deviation	RSD	relative standard deviation
s^2	sample variance		

Special Typefaces

Special typefaces help the reader quickly distinguish letters, words, or phrases from the rest of the text. In your manuscript, underscore (with a simple straight line) all material that is to appear in italics. Mark with a wavy underline material that is to appear in boldface type. Use a double underline to indicate small capital letters. If you are not sure whether material is to be set in a special typeface, do not mark it. A copy editor will mark it appropriately.

Italic Type

- Italic type is used for single letters that denote mathematical constants, variables, and unknown quantities in text and in equations. When such single letters become adjective combinations, they are still italic.

  In this equation, V_i is the frequency of the *i*th mode.

- Italic type is used for genus and species names of all animals, plants, and microorganisms, but not when these names are made plural and adjectival.

  Staphylococcus aureus is a bacterium; staphylococcal infection is what it causes.

- Italic type may be used sparingly to emphasize a word or phrase; it should not be used for long passages.

- Italic type may be used for a word being defined or for a newly introduced word the first time it appears in text.

  In an *outer-sphere transfer*, an electron moves from reductant to oxidant with no chemical alteration of the primary coordination spheres.

- Italic type is used for legal citations, including the "*v.*"

  Rogers v. Massachusetts

- Positional and structural prefixes are set in italic type only when they appear with the chemical name or formula. They are never capitalized, even at the beginning of a sentence or in a title. (The first letter of the chemical name is capitalized at the beginning of a sentence.)

  Examples

  o, m, p, n, sec, tert

  See the section "Chemical Names" for specific compound names.

- Chemical element symbols used to denote attachment to an atom are set in italic type.

  N-ethylaniline   *S*-methyl benzenethiosulfonate

- Configurational prefixes are set in italic type when they appear with the chemical name or formula. In these examples, those that are lower case are never capitalized, even at the beginning of a sentence. Those that are written in capital letters are never lower cased. (The first letter of the chemical name is capitalized at the beginning of a sentence.)

Examples

(R), (S), (Z), (E), *cis, trans, cisoid, transoid, rel, d, l, meso, sn, endo, exo,*

syn, sym, anti, amphi, erythro, threo, altro, ribo, xylo, vic, gem

M (left-handed helix)   *P* (right-handed helix)

- Italic letters within square brackets are used to form names for polycyclic aromatic compounds.

dibenzo[*c,g*]phenanthrene

dibenz[*a,j*]anthracene

indeno[1,2-*a*]indene

dicyclobuta[*de,ij*]naphthalene

1*H*-benzo[*de*]naphthacene

- In polymer nomenclature, *co, alt, b, g, r,* and *m* are set in italic type when they appear with the chemical name or formula. They are never capitalized.

poly(styrene-*co*-butadiene)

poly[(methyl methacrylate)-*co*-styrene]

poly[(methyl methacrylate)-*b*-(styrene-*co*-butadiene)]

poly(ethylene-*alt*-carbon monoxide)

poly(styrene-*g*-acrylonitrile)

- In the newer polymer nomenclature, the words *block, graft, cross, inter,* and *blend* are italicized when they appear with a chemical name. They are never capitalized.

polystyrene-*block*-polybutadiene

polybutadiene-*graft*-[polystyrene:poly(methyl methacrylate)]

poly(*cross*-butadiene)

poly(vinyl trichloroacetate)-*cross*-polystyrene

poly[*cross*-(ethyl acrylate)]-*inter*-polybutadiene

polyisoprene-*blend*-polystyrene

- Italic type is used for symmetry groups and space groups.

C_{2v} $D_{\infty h}$ $I4_1/a$ $Fd3m$ $R\bar{3}_e$

- Italic type is used for certain chirality symbols and symmetry site terms.

Examples

C	clockwise
A	anticlockwise
CU	cube
DD	dodecahedron
TPS	trigonal prism square face tricapped
TP	trigonal pyramid

These are often combined with coordination numbers and position designations to give stereochemical descriptors (e.g., *TP*-6-11′1″).

- Italic type is used for the titles and abbreviations of periodicals, books, and newspapers.

Roman Face (Not Italic)

- Multiplying prefixes are not italic and are always spelled solid (without hyphens or space).

  Latin prefixes: di, tri, tetra, penta, hexa, hepta, octa, nona, deca

  Greek prefixes: bis, tris, tetrakis, pentakis, hexakis, heptakis, octakis, nonakis, decakis

triethyl phosphate	tetrakis(hydroxymethyl)methane
hexachlorobenzene	tris(ethylenediamine)cadmium
1,4-bis(3-bromo-1-	dihydroxide
oxopropyl)piperazine	

The first letter of these words is capitalized as the first word of a sentence and in titles and headings.

- Chemical formulas are not italic.

  H_3PO_4 C_6H_5OH HCl   NaOH $BaSO_4$

- Italic type is not used for common Latin terms and abbreviations:

ab initio	et al.	in vivo
ad hoc	etc.	status quo
a priori	i.e.	vs.
ca.	in situ	
e.g.	in vitro	

- Structural prefixes "cyclo" and "iso" are set in Roman type and are spelled solid (without hyphens or space).

    cyclohexane   isopropyl alcohol

    Exception: *cyclo*-hexasulfur

They are capitalized as the first word of a sentence and in titles and headings.

- Italic type is not used for "pH"; "p" is always lower case and "H" is always capitalized.

- Italic type is not used for chemical concentration unit M (molar, mol/dm^3, mol/L), but it is necessary for m (molal, mol/kg).

    1.0 M HCl   *but*   2.0 m NaOH

Greek Letters

- Greek letters, not the spelled-out forms, are used in chemical and drug names.

    α-amino acid   *not*   alpha amino acid

    β-naphthol   *not*   beta naphthol

When such names are the first word of a sentence or when they appear in titles or headings, the Greek letter is retained and the first non-Greek Roman letter is capitalized.

    α-Ethylbenzeneethanol is an oily liquid with a mild odor.

- Greek letters are used for some bonding orbitals.

    π bond σ orbital

Small Capital Letters

- Small capital letters D and L are used for absolute configuration with amino acids and carbohydrates.

    D-glucose   L-alanine

Boldface Type

- Boldface Roman type is used for vectors and tensors.

- Boldface Roman type is used for Arabic compound numbers.

Units of Measure

- Use metric and SI units (see next section).

- Abbreviate the unit of measure when it appears with a numeral; leave a space between a number and the unit of measure, except when they form a unit modifier, in which case use a hyphen between them.

    500 mL of NaOH   *but*   500-mL flask

- Spell out the unit of measure when no quantity is given.

    several milligrams   *not*   several mg

    a few milliliters   *not*   a few mL

- Do not use plurals for abbreviated units of measure.

    50 mg   *not*   50 mgs

- In ranges and series, retain only the final unit of measure.

    10–12 mg   5, 10, and 20 kV   25–30%

    between 25 and 50 mL   from 10 to 15 min

    Exception

    60° and 90° rotations

- When a sentence starts with a specific quantity, spell it out as well as the unit of measure.

    Twenty-five milliliters of acetone was added, and the mixture was centrifuged.

    However, it is preferable to recast the sentence:

    Acetone (25 mL) was added, and the mixture was centrifuged.

- Even when a sentence starts with a spelled-out quantity, use numerals when appropriate in the rest of the sentence.

  Twenty-five milliliters of acetone and 5 mL of HCl were added.

- Use the percent symbol with a number, without a space.

  25%   45–50%

- Use the Celsius or kelvin rather than the Fahrenheit temperature scale.

- Do not use the degree symbol with kelvin: 115 K.

- Use °C with a space after a number, but no space between the degree symbol and the capital C: 15 °C.

- For angle measurements, use the degree symbol closed up to the number: 12°.

- Certain units of measure have no abbreviations: bar, einstein, torr.

The International System of Units

Before the 1960s, four systems of units were commonly used in the scientific literature: the English system (centuries old, using yard and pound), the metric system (dating from the 18th century, using meter and kilogram as standard units), the CGS system (based on the metric, using centimeter, gram, and second as base units), and the MKSA or Giorgi system (using meter, kilogram, second, and ampere as base units).

The International System of Units (SI, Système International) is the most recent effort to develop a coherent system of units. It is coherent because there is only one unit for each base physical quantity, and units for all other quantities are derived from these base units by simple equations. It is coherent also in the sense that it has been adopted as a universal system to facilitate communication of numerical data and to restrict proliferation of systems.

SI is constructed from seven base units for independent quantities plus two supplementary units for plane and solid angles: meter, kilogram, second, ampere, kelvin, mole, candela, radian, and steradian. Most physicochemical measurements can be expressed in terms of these units.

Certain units not part of SI are so widely used that it is impractical to abandon them (e.g., liter, minute, and hour) or so well established that the

International Committee on Weights and Measures has authorized their continued use (e.g., bar, curie, and angstrom). These exceptions are permitted in ACS books and journals. In addition, quantities that are expressed in terms of the fundamental constants of nature, such as elementary charge, proton mass, Bohr magneton, speed of light, and Planck constant, are also acceptable. However, broad terms such as "atomic units" are not acceptable, although atomic mass unit, u, is acceptable and relevant to chemistry. See the Bibliography for references on the SI.

- Do not capitalize SI units when they are spelled out, except in titles.

- Use abbreviations for units only with numbers. Otherwise, spell them out.

- With numbers, use the abbreviations for SI units with capital and lower-case letters exactly as they appear in the following tables.

- Add an "s" to form the plural of SI units when they are spelled out but not when they are abbreviated. Abbreviated units do not have plurals.

- Do not use a period after an abbreviated unit.

- Leave a full space between the number and the unit, except as a unit modifier, in which case use a hyphen.

- Leave no space between the multiplicative prefix and the unit, whether abbreviated or spelled out.

    kilojoule or kJ   milligram or mg

- For abbreviated compound units, use a center dot for multiplication and a solidus or negative exponent for division.

    watt per meter kelvin is $W/(m \cdot K)$, $W \cdot (m \cdot K)^{-1}$, or $W\ m^{-1}\ K^{-1}$

- Do not mix abbreviations and words.

    Results are shown in newtons per meter not N per meter

    100 F/m not 100 farad/m

Table I. SI Units

Unit	Symbol	Physical Quantity
Base units		
ampere	A	electric current
candela	cd	luminous intensity
kelvin	K	thermodynamic temperature
kilogram	kg	mass
meter	m	length
mole	mol	amount of substance
second	s	time
Supplementary units		
radian	rad	plane angle
steradian	sr	solid angle

Table II. Multiplying Prefixes

Factor	Prefix	Symbol	Factor	Prefix	Symbol
10^{-18}	atto	a	10^{1}	deka	da
10^{-15}	femto	f	10^{2}	hecto	h
10^{-12}	pico	p	10^{3}	kilo	k
10^{-9}	nano	n	10^{6}	mega	M
10^{-6}	micro	μ	10^{9}	giga	G
10^{-3}	milli	m	10^{12}	tera	T
10^{-2}	centi	c	10^{15}	peta	P
10^{-1}	deci	d	10^{18}	exa	E

NOTE: Any of these prefixes may be combined with any of the symbols permitted within SI. Thus, kPa and GPa will both be common combinations in measurements of pressure, as will mL and cm for measurements of volume and length. As a general rule, however, the prefix chosen should be 10 raised to that multiple of 3 that will bring the numerical value of the quantity to a positive value less than 1000.

Table III. SI-Derived Units

Unit	Symbol	Quantity	In Terms of Other Units	In Terms of SI Base Units
becquerel	Bq	activity (of a radionuclide)		s^{-1}
coulomb	C	quantity of electricity, electric charge		$A \cdot s$, $s \cdot A$
farad	F	capacitance	C/V	$m^{-2} \cdot kg^{-1} \cdot s^{4} \cdot A^{2}$
gray	Gy	absorbed dose, kerma, specific energy imparted	J/kg	$m^{2} \cdot s^{-2}$
henry	H	inductance	Wb/A	$m^{2} \cdot kg \cdot s^{-2} \cdot A^{-2}$
hertz	Hz	frequency		s^{-1}
joule	J	energy, work, quantity of heat	N·m	$m^{2} \cdot kg \cdot s^{-2}$
lumen	lm	luminous flux	lm	$cd \cdot sr$
lux	lx	illuminance	lm/m^{2}	$m^{-2} \cdot cd \cdot sr$
newton	N	force		$m \cdot kg \cdot s^{-2}$
ohm	Ω	electric resistance	V/A	$m^{2} \cdot kg \cdot s^{-3} \cdot A^{-2}$
pascal	Pa	pressure, stress	N/m^{2}	$m^{-1} \cdot kg \cdot s^{-2}$
siemens	S	conductance	A/V	$m^{-2} \cdot kg^{-1} \cdot s^{3} \cdot A^{2}$
sievert	Sv	dose equivalent	J/kg	$m^{2} \cdot s^{-2}$
tesla	T	magnetic flux density	Wb/m^{2}	$kg \cdot s^{-2} \cdot A^{-1}$
volt	V	electric potential, potential difference, electromotive force	W/A	$m^{2} \cdot kg \cdot s^{-3} \cdot A^{-1}$
watt	W	power, radiant flux	J/s	$m^{2} \cdot kg \cdot s^{-3}$
weber	Wb	magnetic flux	V·s	$m^{2} \cdot kg \cdot s^{-2} \cdot A^{-1}$

Table IV. SI-Derived Compound Units

Unit	Symbol	Quantity	In Terms of Other Units
ampere per meter	A/m	magnetic field strength	
ampere per square meter	A/m²	current density	
candela per square meter	cd/m²	luminance	
coulomb per cubic meter	C/m³	electric charge density	$m^{-3} \cdot s \cdot A$
coulomb per kilogram	C/kg	exposure (X- and γ-rays)	
coulomb per square meter	C/m²	electric flux density	$m^{-2} \cdot s \cdot A$
cubic meter	m³	volume	
cubic meter per kilogram	m³/kg	specific volume	
farad per meter	F/m	permittivity	$m^{-3} \cdot kg^{-1} \cdot s^4 \cdot A^2$
henry per meter	H/m	permeability	$m \cdot kg \cdot s^{-2} \cdot A^{-2}$
joule per cubic meter	J/m³	energy density	$m^{-1} \cdot kg \cdot s^{-2}$
joule per kelvin	J/K	heat capacity, entropy	$m^2 \cdot kg \cdot s^{-2} \cdot K^{-1}$
joule per kilogram	J/kg	specific energy	$m^2 \cdot s^{-2}$
joule per kilogram kelvin	J/(kg·K)	specific heat capacity, specific entropy	$m^2 \cdot s^{-2} \cdot K^{-1}$
joule per mole	J/mol	molar energy	$m^2 \cdot kg \cdot s^{-2} \cdot mol^{-1}$
joule per mole kelvin	J/(mol·K)	molar entropy, molar heat capacity	$m^2 \cdot kg \cdot s^{-2} \cdot K^{-1} \cdot mol^{-1}$
kilogram per cubic meter	kg/m³	density, mass density	
meter per second	m/s	speed, velocity	
meter per square second	m/s²	acceleration	
mole per cubic meter[a]	mol/m³	concentration (amount of substance)	
newton meter	N·m	moment of force	$m^2 \cdot kg \cdot s^{-2}$
newton per meter	N/m	surface tension	$kg \cdot s^{-2}$
pascal second	Pa·s	dynamic viscosity	$m^{-1} \cdot kg \cdot s^{-1}$

Continued on next page

Table IV. Continued

Unit	Symbol	Quantity	In Terms of Other Units
radian per second	rad/s	angular velocity	
radian per square second	rad/s^2	angular acceleration	
reciprocal meter	m^{-1}	wavenumber	
reciprocal second	s^{-1}	frequency	
square meter	m^2	area	
square meter per second	m^2/s	kinematic viscosity	
volt per meter	V/m	electric field strength	m·kg·s^{-3}·A^{-1}
watt per meter kelvin	W/(m·K)	thermal conductivity	m·kg·s^{-3}·K^{-1}
watt per square meter	W/m^2	heat flux density, irradiance	kg·s^{-3}
watt per square meter steradian	W/(m^2·sr)	radiance	
watt per steradian	W/sr	radiant intensity	

[a] Liter (L) is a special name for cubic decimeter. The symbol M should not be used as a unit, but expressions such as 0.1 M, meaning a solution with concentration of 0.1 mol/L, are acceptable.

Table V. Other Units in Use with the SI

Unit	Symbol	Quantity	Value in SI Units
angstrom	Å	distance	$1\ \text{Å} = 10^{-10}\ \text{m} = 0.1\ \text{nm}$
bar	bar	pressure	$1\ \text{bar} = 10^5\ \text{Pa} = 0.1\ \text{MPa}$
barn	b	cross section	$1\ \text{b} = 10^{-28}\ \text{m} = 100\ \text{fm}^2$
curie	Ci	activity	$1\ \text{Ci} = 3.7 \times 10^{10}\ \text{Bq}$
day	day	time	$1\ \text{day} = 24\ \text{h} = 86\,400\ \text{s}$
degree	°	plane angle	$1° = (\pi/180)\ \text{rad}$
degree Celsius	°C	temperature	$—^a$
electronvolt	eV	$—^b$	$1\ \text{eV} = 1.602\,19 \times 10^{-19}\ \text{J}$
hectare	ha	area	$1\ \text{ha} = 1\ \text{hm}^2 = 10^4\ \text{m}^2$
hour	h	time	$1\ \text{h} = 60\ \text{min} = 3600\ \text{s}$
liter	L	volume	$1\ \text{L} = 1\ \text{dm}^3 = 10^{-3}\ \text{m}^3$
metric ton	t	mass	$1\ \text{t} = 10^3\ \text{kg}$
minute	min	time	$1\ \text{min} = 60\ \text{s}$
minute	′	plane angle	$1′ = (1/60)° = (\pi/10\,800)\ \text{rad}$
rad	rad^c	absorbed dose	$1\ \text{rad} = 0.01\ \text{Gy} = 1\ \text{cGy}$
roentgen	R	exposure	$1\ \text{R} = 2.58 \times 10^{-4}\ \text{C/kg}$
second	″	plane angle	$1″ = (1/60)′ = (\pi/648\,000)$ rad
unified atomic mass unit	u	$—^d$	$1\ \text{u} = 1.660\,57 \times 10^{-27}\ \text{kg}$

[a] Temperature intervals in kelvins and degrees Celsius are identical; however, temperature in kelvins equals temperature in degrees Celsius plus 273.15.

[b] The electronvolt is the kinetic energy acquired by an electron in passing through a potential difference of 1 V in vacuum.

[c] When there is a possibility of confusion with the symbol for radian, rd may be used as the symbol for rad.

[d] The unified atomic mass unit is equal to (1/12) of the mass of an atom of the nuclide ^{12}C.

Table VI. Non-SI Units That Are Discouraged

Unit	Value in SI Units
kilogram-force	9.80665 N
calorie (thermochemical)	4.184 J
mho	1 S
standard atmosphere	101.325 kPa
technical atmosphere	98.0665 kPa
conventional millimeter of mercury	133.322 Pa
torr	133.322 Pa
grad	$2\pi/400$ rad
metric carat	0.2 g
metric horsepower	735.499 W
micron	$1\ \mu\text{m}$

Chemical Names

Chemical compounds should be named according to the rules established by the appropriate international nomenclature organization. Those rules will not be presented here; the Bibliography lists nomenclature references. This section presents editorial conventions and style points for typesetting chemical names that are presumed to be correct.

The names of chemical compounds consist of locants, descriptors, and syllabic portions. Locants and descriptors can be numbers, element symbols, small capital letters, Greek letters, Roman letters, italic words and letters, and combinations of these. The syllabic portions of chemical names are just like other common nouns: they are set in Roman face, they are lower case in text, they are capitalized at the beginnings of sentences, and they are hyphenated only when they do not fit completely on one line. Tables VII and VIII present examples of chemical names and their capitalization.

Tables VII and VIII illustrate many of the following points.

- Commas are used between numerical locants, chemical element symbol locants, and Greek locants, without a space after the comma. However, some single locants consist of a number and a Greek letter together with no space or punctuation; in these cases, the number precedes the Greek letter. When the Greek letter precedes the number, they are two different locants and should be separated by a comma. For example, $\alpha,2$ are two locants; 1α is one locant.

- Hyphens are used to separate locants and configurational descriptors from the syllabic portion of the name. Hyphens are not placed in chemical names if they are not separating a locant or descriptor from the syllabic portion. Hyphens are not used to separate the syllables of a chemical name, unless the name is too long to fit on one line. See the next section, "Recommended End-of-Line Hyphenation of Chemical Names".

- Chemical names and nonproprietary drug names are not capitalized unless they are the first word of a sentence or in a title or heading. Then, the first letter of the syllabic portion is capitalized, not the locant, stereoisomer descriptor, or positional prefix.

- Positional prefixes are set in italic type when they appear with the chemical name or formula.

- Chemical element symbols used to denote attachment to an atom are set in italic type.

Table VII. Examples of Chemical Names That Usually Consist of More Than One Word

In Text	At Beginning of Sentence	In Titles, Headings
Acids		
benzoic acid	Benzoic acid	Benzoic Acid
hydrochloric acid	Hydrochloric acid	Hydrochloric Acid
ethanethioic *S*-acid	Ethanethioic *S*-acid	Ethanethioic *S*-Acid
Alcohols		
ethyl alcohol	Ethyl alcohol	Ethyl Alcohol
ethylene glycol	Ethylene glycol	Ethylene Glycol
Ketones		
methyl phenyl ketone	Methyl phenyl ketone	Methyl Phenyl Ketone
di-2-naphthyl ketone	Di-2-naphthyl ketone	Di-2-naphthyl Ketone
Ethers		
methyl propyl ether	Methyl propyl ether	Methyl Propyl Ether
di-*sec*-butyl ether	Di-*sec*-butyl ether	Di-*sec*-butyl Ether
Anhydrides		
acetic anhydride	Acetic anhydride	Acetic Anhydride
phthalic anhydride	Phthalic anhydride	Phthalic Anhydride
Esters		
methyl acetate	Methyl acetate	Methyl Acetate
phenyl thiocyanate	Phenyl thiocyanate	Phenyl Thiocyanate
propyl benzoate	Propyl benzoate	Propyl Benzoate
lithium propanoate	Lithium propanoate	Lithium Propanoate
Salts		
butyl chloride	Butyl chloride	Butyl Chloride
aniline hydrochloride	Aniline hydrochloride	Aniline Hydrochloride
2-naphthoyl bromide	2-Naphthoyl bromide	2-Naphthoyl Bromide
tert-butyl fluoride	*tert*-Butyl fluoride	*tert*-Butyl Fluoride
methyl iodide	Methyl iodide	Methyl Iodide
magnesium oxide	Magnesium oxide	Magnesium Oxide
ammonium hydroxide	Ammonium hydroxide	Ammonium Hydroxide
dicyclohexyl peroxide	Dicyclohexyl peroxide	Dicyclohexyl Peroxide
benzyl hydroperoxide	Benzyl hydroperoxide	Benzyl Hydroperoxide
diethyl sulfide	Diethyl sulfide	Diethyl Sulfide
diethyl disulfide	Diethyl disulfide	Diethyl Disulfide
dimethyl sulfate	Dimethyl sulfate	Dimethyl Sulfate
sodium *S*-phenyl thiosulfite	Sodium *S*-phenyl thiosulfite	Sodium *S*-Phenyl Thiosulfite
sodium cyanide	Sodium cyanide	Sodium Cyanide

Table VIII. Locants and Descriptors in Chemical Names

In Text	At Beginning of Sentence	In Titles, Headings
Number Locants		
2-benzoylbenzoic acid	2-Benzoylbenzoic acid	2-Benzoylbenzoic Acid
1,2-dicyanobutane	1,2-Dicyanobutane	1,2-Dicyanobutane
1-bromo-3-chloropropane	1-Bromo-3-chloropropane	1-Bromo-3-chloropropane
adenosine 5′-triphosphate	Adenosine 5′-triphosphate	Adenosine 5′-Triphosphate
4a,8a-dihydronaphthalene	4a,8a-Dihydronaphthalene	4a,8a-Dihydronaphthalene
2-(2-chloroethyl)pentanoic acid	2-(2-Chloroethyl)pentanoic acid	2-(2-Chloroethyl)pentanoic Acid
7-(4-chlorophenyl)-1-naphthol	7-(4-Chlorophenyl)-1-naphthol	7-(4-Chlorophenyl)-1-naphthol
Element Symbol Locants		
N-ethylaniline	N-Ethylaniline	N-Ethylaniline
N,N′-dimethylurea	N,N′-Dimethylurea	N,N′-Dimethylurea
S-methyl benzenethiosulfonate	S-Methyl benzenethiosulfonate	S-Methyl Benzenethiosulfonate
O,O,S-triethyl phosphorodithioate	O,O,S-Triethyl phosphorodithioate	O,O,S-Triethyl Phosphorodithioate
3H-fluorene	3H-Fluorene	3H-Fluorene
N,2-dihydroxybenzamide	N,2-Dihydroxybenzamide	N,2-Dihydroxybenzamide
Greek Letter Locants		
α-methylbenzeneacetic acid	α-Methylbenzeneacetic acid	α-Methylbenzeneacetic Acid
α₁-sitosterol	α₁-Sitosterol	α₁-Sitosterol
β-chloro-1-naphthalenebutanol	β-Chloro-1-naphthalenebutanol	β-Chloro-1-naphthalenebutanol
β-endorphin	β-Endorphin	β-Endorphin
α-hydroxy-β-aminobutyric acid	α-Hydroxy-β-aminobutyric acid	α-Hydroxy-β-aminobutyric Acid
1α-hydroxycholecalciferol	1α-Hydroxycholecalciferol	1α-Hydroxycholecalciferol
γ-hydroxy-β-aminobutyric acid	γ-Hydroxy-β-aminobutyric acid	γ-Hydroxy-β-aminobutyric Acid
17α-hydroxy-5β-pregnane	17α-Hydroxy-5β-pregnane	17α-Hydroxy-5β-pregnane
ω,ω′-dibromopolybutadiene	ω,ω′-Dibromopolybutadiene	ω,ω′-Dibromopolybutadiene
β,4-dichlorocyclohexanepropionic acid	β,4-Dichlorocyclohexanepropionic acid	β,4-Dichlorocyclohexanepropionic Acid

Small Capital Letter Locants		
L-methionine	L-Methionine	L-Methionine
D-serine	D-Serine	D-Serine
DL-alanine	DL-Alanine	DL-Alanine
β-D-arabinose	**β**-D-Arabinose	**β**-D-Arabinose
D-1,2,4-butanetriol	D-1,2,4-Butanetriol	D-1,2,4-Butanetriol
D_S-threonine	D_S-Threonine	D_S-Threonine
Positional Descriptors[a]		
cis-1,2-dichloroethene	*cis*-1,2-Dichloroethene	*cis*-1,2-Dichloroethene
trans-1,2-dideuteriocyclobutane	*trans*-1,2-Dideuteriocyclobutane	*trans*-1,2-Dideuteriocyclobutane
1-(*trans*-1-propenyl)-3-(*cis*-1-propenyl)naphthalene	1-(*trans*-1-Propenyl)-3-(*cis*-1-propenyl)naphthalene	1-(*trans*-1-Propenyl)-3-(*cis*-1-propenyl)naphthalene
o-dibromobenzene	*o*-Dibromobenzene	*o*-Dibromobenzene
m-hydroxybenzyl alcohol	*m*-Hydroxybenzyl alcohol	*m*-Hydroxybenzyl Alcohol
p-benzenediacetic acid	*p*-Benzenediacetic acid	*p*-Benzenediacetic Acid
2-(*o*-chlorophenyl)-1-naphthol	2-(*o*-Chlorophenyl)-1-naphthol	2-(*o*-Chlorophenyl)-1-naphthol
4-chloro-*m*-cresol	4-Chloro-*m*-cresol	4-Chloro-*m*-cresol
7-bromo-*p*-cymene	7-Bromo-*p*-cymene	7-Bromo-*p*-cymene
n-butyl iodide	*n*-Butyl iodide	*n*-Butyl Iodide
sec-butyl alcohol	*sec*-Butyl alcohol	*sec*-Butyl Alcohol
tert-amyl isovalerate	*tert*-Amyl isovalerate	*tert*-Amyl Isovalerate
p-*tert*-butylphenol	*p*-*tert*-Butylphenol	*p*-*tert*-Butylphenol
trans-cisoid-trans-perhydrophenanthrene	*trans-cisoid-trans*-Perhydrophenanthrene	*trans-cisoid-trans*-Perhydrophenanthrene
s-triazine	*s*-Triazine	*s*-Triazine
sym-dibromoethane	*sym*-Dibromoethane	*sym*-Dibromoethane
Stereoisomer Descriptors[b]		
(*E*)-diphenyldiazene	(*E*)-Diphenyldiazene	(*E*)-Diphenyldiazene
(*Z*)-5-chloro-4-pentenoic acid	(*Z*)-5-Chloro-4-pentenoic acid	(*Z*)-5-Chloro-4-pentenoic Acid
(1*Z*,4*E*)-1,2,4,5-tetrachloro-1,4-pentadiene	(1*Z*,4*E*)-1,2,4,5-Tetrachloro-1,4-pentadiene	(1*Z*,4*E*)-1,2,4,5-Tetrachloro-1,4-pentadiene

Continued on next page

Table VIII. Continued

In Text	At Beginning of Sentence	In Titles, Headings
(S)-2,3-dihydroxypropanoic acid	(S)-2,3-Dihydroxypropanoic acid	(S)-2,3-Dihydroxypropanoic Acid
(1R*,3S*)-1-bromo-3-chlorocyclohexane	(1R*,3S*)-1-Bromo-3-chlorocyclohexane	(1R*,3S*)-1-Bromo-3-chlorocyclohexane
rel-(1R,3R)-1-bromo-3-chlorocyclohexane	rel-(1R,3R)-1-Bromo-3-chlorocyclohexane	rel-(1R,3R)-1-Bromo-3-chlorocyclohexane
d-camphor	d-Camphor	d-Camphor
dl-2-aminopropanoic acid	dl-2-Aminopropanoic acid	dl-2-Aminopropanoic Acid
s-triazine	s-Triazine	s-Triazine
meso-tartaric acid	meso-Tartaric acid	meso-Tartaric Acid
sn-glycerol 1-(dihydrogen phosphate)	sn-Glycerol 1-(dihydrogen phosphate)	sn-Glycerol 1-(Dihydrogen Phosphate)
endo-2-chlorobicyclo[2.2.1]heptane	endo-2-Chlorobicyclo[2.2.1]heptane	endo-2-Chlorobicyclo[2.2.1]heptane
exo-bicyclo[2.2.2]oct-5-en-2-ol	exo-Bicyclo[2.2.2]oct-5-en-2-ol	exo-Bicyclo[2.2.2]oct-5-en-2-ol
exo-5,6-dimethyl-endo-bicyclo[2.2.2]octan-2-ol	exo-5,6-Dimethyl-endo-bicyclo[2.2.2]octan-2-ol	exo-5,6-Dimethyl-endo-bicyclo[2.2.2]octan-2-ol
syn-7-methylbicyclo[2.2.1]heptene	syn-7-Methylbicyclo[2.2.1]heptene	syn-7-Methylbicyclo[2.2.1]heptene
sym-dibromoethane	sym-Dibromoethane	sym-Dibromoethane
anti-bicyclo[3.2.1]octan-8-amine	anti-Bicyclo[3.2.1]octan-8-amine	anti-Bicyclo[3.2.1]octan-8-amine
erythro-β-hydroxyaspartic acid	erythro-β-Hydroxyaspartic acid	erythro-β-Hydroxyaspartic Acid
(+)-erythro-2-amino-3-methylpentanoic acid	(+)-erythro-2-Amino-3-methylpentanoic acid	(+)-erythro-2-Amino-3-methylpentanoic Acid
threo-2-amino-1-hydroxy-1-phenylpropane	threo-2-Amino-1-hydroxy-1-phenylpropane	threo-2-Amino-1-hydroxy-1-phenylpropane
L-threo-2,3-dichlorobutyric acid	L-threo-2,3-Dichlorobutyric acid	L-threo-2,3-Dichlorobutyric Acid
trans-cisoid-trans-perhydrophenanthrene	trans-cisoid-trans-Perhydrophenanthrene	trans-cisoid-trans-Perhydrophenanthrene

[a] Other positional descriptors are vic and gem.

[b] Stereoisomer descriptors include (R), (S), (Z), (E), cis, trans, cisoid, transoid, rel, d, l, meso, sn, endo, exo, syn, anti, amphi, erythro, threo, altro, ribo, xylo; M (left-handed helix); and P (right-handed helix).

- Stereoisomer descriptors are set in italic type when they appear with the chemical name or formula.

- Multiplying prefixes are set in Roman type and closed up to the word.

  Latin prefixes
  di, tri, tetra, penta, hexa, hepta, octa, nona, deca

  Greek prefixes
  bis, tris, tetrakis, pentakis, hexakis, heptakis, octakis, nonakis, decakis

  Examples

  dichloride   triamine   2,3,5-triaziridin-1-yl-*p*-benzoquinone

  1,3-bis(diethylamino)propane   2,3,5-tris(1-aziridinyl)-*p*-benzoquinone

- In polymer nomenclature, *co*, *alt*, *b*, *g*, *r*, and *m* are set in italic type when they appear with the chemical name or formula. They are never capitalized. At the beginning of a sentence, in titles, and in headings, only the "p" of the examples that follow or the first nonpositional character is capitalized.

  poly(styrene-*co*-butadiene)

  poly[(methyl methacrylate)-*co*-styrene]

  poly(ethylene-*alt*-carbon monoxide)

  poly(styrene-*g*-acrylonitrile)

  poly[(methyl methacrylate)-*b*-(styrene-*co*-butadiene)]

- In the newer polymer nomenclature, the words *block*, *graft*, *cross*, *inter*, and *blend* are italicized when they appear with a chemical name. They are never capitalized. At the beginning of a sentence, in titles, and in headings, only the "p" of the examples that follow or the first nonpositional character is capitalized.

  polystyrene-*block*-polybutadiene

  poly(*cross*-butadiene)

  polyisoprene-*blend*-polystyrene

  poly(vinyl trichloroacetate)-*cross*-polystyrene

  poly[*cross*-(ethyl acrylate)]-*inter*-polybutadiene

  polybutadiene-*graft*-[polystyrene:poly(methyl methacrylate)]

- Greek letters, not the spelled-out forms, are used in chemical and drug names.

    α-amino acid   *not*   alpha amino acid

    β-naphthol   *not*   beta naphthol

- Numbers separated by periods within square brackets are used in names of bridged and spiro alicyclic compounds.

    bicyclo[3.2.0]heptane   spiro[4.5]decane

- Italic letters within square brackets are used in names of polycyclic aromatic compounds.

    dibenz[*a,j*]anthracene   indeno[1,2-*a*]indene
    dicyclobuta[*de,ij*]naphthalene   1*H*-benz[*de*]naphthacene

- Optical rotational signs are indicated by plus and minus signs in parentheses and hyphenated to the chemical name.

    (+)-glucose   (–)-tartaric acid

- Trade names are spelled with an initial capital letter. The trademark symbol is omitted.

End-of-Line Hyphenation of Chemical Names

The following is a list of prefixes, suffixes, roots, and some complete words. Here they are hyphenated as they would be at the end of a line. If, for example, the name "5-(2-chloroethyl)-9-(diaminomethyl)-2-anthracenol" needed to be hyphenated because it would not all fit on the same line, you can look up each syllable in the list for the proper place to hyphenate. Also, chemical names can be broken at the hyphens already included in their names. Follow other standard rules for hyphenation of regular words; for example, leave at least four characters on each line. Thus, the example given could be hyphenated as follows:

5-(2- chlo- ro- eth- yl)- 9- (di- ami- no- meth- yl)- 2- anth- ra- cenol

acet-am-ide	ace-to	acry-lo
acet-am-i-do	ace-tyl	ad-i-po-yl
ace-naph-tho	acryl-ic	al-kyl
ace-tic	acry-late	al-lyl

ami-di-no
amide
ami-do
amine
ami-no
am-mine
am-mo-nio
am-mo-ni-um
an-thra
an-thra-cene
an-thra-cenol
an-thryl
ar-se-nate
ar-si-no
aryl
azi-do
azi-no
azo
azyl
benz-ami-do
ben-zene
benz-hy-dryl
ben-zo-yl
ben-zyl
ben-zyl-i-dene
bi-cy-clo
bo-ryl
bro-mide
bro-mo
bu-tane
bu-ten-yl
bu-tyl
bu-tyl-ene
bu-tyl-i-dene
car-ba-mate
car-ba-mide
car-ba-mi-doyl
car-ba-moyl
car-bon-yl
car-box-a-mi-do
car-box-y
car-box-yl
car-byl-a-mi-no
chlo-ride

chlo-ro
chlo-ro-syl
chlo-ryl
cu-mene
cy-a-nate
cy-a-nide
cy-a-na-to
cy-a-no
cy-clo
cy-clo-hex-ane
cy-clo-hex-yl
di-azo
di-bo-ran-yl
di-car-bon-yl
di-im-ino
di-oxy
di-oyl
diyl
ep-oxy
eth-ane
eth-a-nol
eth-a-noyl
eth-enyl
eth-yl
eth-yl-ene
eth-yl-i-dene
eth-ynyl
fluo-ride
fluo-ro
form-al-de-hyde
form-ami-do
for-mic
form-imi-doyl
for-myl
fu-ran
ger-myl
guan-i-di-no
gua-nyl
halo
hep-tane
hep-tyl
hex-ane
hex-yl
hy-dra-zi-no

hy-dra-zo
hy-dro
hy-drox-ide
hy-droxy
imi-da-zole
imide
imi-do
imi-doyl
imi-no
in-da-mine
in-da-zole
in-dene
in-de-no
in-dole
io-date
io-dide
iodo
io-do-syl
io-dyl
iso-cy-a-na-to
iso-cy-a-nate
iso-cy-a-nide
iso-pro-pen-yl
iso-pro-pyl
mer-cap-to
mer-cu-ric
meth-an-ami-do
meth-ane
meth-anol
meth-an-oyl
meth-yl
meth-yl-ate
meth-yl-ene
meth-yl-i-dene
naph-tha-lene
naph-tho
naph-thyl
neo-pen-tyl
ni-trate
ni-tric
ni-trile
ni-trilo
ni-trite
ni-tro

ni-troso
oc-tane
oc-tyl
ox-idase
ox-ide
ox-ido
ox-ime
oxo
ox-o-nio
oxy
pen-tane
pen-tyl
pen-tyl-i-dene
per-chlo-rate
per-chlo-ride
per-chlo-ryl
per-man-ga-nate
per-ox-idase
per-ox-ide
per-oxy
phe-nan-threne
phe-nan-thro
phe-nan-thryl
phe-nol
phen-yl
phen-yl-ene
phos-phate
phos-phide
phos-phine
phos-phino
phos-phin-yl
phos-pho
phos-pho-nio
phos-pho-no
phos-phor-ic
phos-phor-anyl
plum-byl
pro-pane
pro-pen-yl
pro-pen-yl-ene
pro-pyl
pro-pyl-ene
pro-pyl-i-dene

pu-rine
py-ran
pyr-a-zine
pyr-a-zole
pyr-i-dine
pyr-id-a-zine
pyr-role
quin-o-line
qui-none
se-le-nic
se-le-no
se-le-nite
si-lane
sil-anyl
sil-ox-anyl
sil-ox-y
si-lyl
spi-ro
stan-nic
stan-nite
stan-nous
stan-nyl
stib-ino
sty-rene
sty-ryl
sul-fa-moyl
sul-fate
sul-fen-a-moyl
sul-fe-no
sul-fe-nyl
sul-fide
sul-fi-do
sul-fin-a-moyl
sul-fi-no
sul-fi-nyl
sul-fite
sul-fo
sul-fon-ami-do
sul-fo-nate
sul-fone
sul-fon-ic
sul-fo-nio
sul-fo-nyl

sulf-ox-ide
sul-fu-rate
sul-fu-ric
sul-fu-rous
sul-fu-ryl
thio
thio-nyl
thio-phene
thi-oxo
thi-oyl
tol-u-ene
tol-u-ide
tol-yl
tri-a-zine
tri-a-zole
tri-yl
urea
ure-ide
ure-id-o
uric
vi-nyl
vi-nyl-i-dene
xan-thene
xan-tho
xy-lene
xy-li-dine
xy-lyl
xy-li-din-yl
yl-i-dene

Chemical Elements, Symbols, Nuclides, and Formulas

- Both chemical symbols and element names may be used in text, but they should be used consistently, not mixed.

  NaCl   *not*   Na chloride

  However, NaCl and sodium chloride are both acceptable within the same paper.

- Symbols for chemical elements are written in Roman type with an initial capital letter as individual atoms and as parts of formulas.

  Ca   C   H   He   HCl   NaOH $HgSO_4$

- The names of the chemical elements and formulas are also written in Roman type, but they are treated as common nouns.

  calcium   carbon   hydrogen   helium

  hydrochloric acid   sodium hydroxide   mercuric sulfate

- Nuclides may be specified by attaching numbers to the element symbol in the following positions: mass number, left superscript; atomic number, left subscript; ionic charge, state of excitation, or oxidation number, right superscript; and number of atoms per molecule, right subscript.

  mass number: ^{32}S ^{12}C ^{35}Cl

  atomic number: $_6C$ $_{16}S$

  ionic charge: Na^+ NO_3^- Ca^{2+} PO_4^{3-} $_7Li^-$

  excited electronic state: He^* NO^*

  oxidation number: Pb_2^{II} $Pb^{IV}O_2$ $(NH_3)_2Pt^{II}$

  number of atoms: C_6 Fe_3 NH_4

- In narrative text and not in formulas, oxidation numbers may also be indicated by a Roman number on the line and in parentheses.

formula	*text*
Co^{III}	cobalt(III) *or* Co(III) *or* Co^{III}
$Fe^{II}Cl_2$	iron(II) chloride
$(NH_3)_2Pt^{II}$	diammineplatinum(II)
$Mn^{IV}O_2$	manganese(IV) oxide

- To designate ionic charge, multiple plus or minus signs are not used, the number is written to the left of the charge, and the charge is not circled.

Cl^- *not* $Cl^{\ -}$

Hg^{2+} *not* Hg^{++} *not* Hg^{+2}

- In narrative text, words or symbols and numbers on the line and hyphenated may be used to refer to an atom in a specific position.

at the carbon in the 6-position   *or*   at C-6

- In the formula of a free radical, the unshared electron is indicated by a centered dot.

$H_3C\cdot$ $C_6H_5\cdot$ $HO\cdot$

- In the formula for an addition compound, a centered dot is also used.

$Na_2SO_4\cdot 10H_2O$ $BH_3\cdot NH_3$

- Cation and anion radicals are indicated by the symbol, formula, or structure with a superscript dot followed by a plus or minus sign. Plus and minus signs are not enclosed in circles.

$R^{\cdot+}$ $R_2^{\cdot+}$ $R^{\cdot-}$ $R^{\cdot 2-}$ $C_6H_5NO^{\cdot 3-}$

- Bonds are indicated by dashes, not by dots, and only when necessary. For linear formulas in text, no single bonds are necessary.

H_2SO_4 $C_6H_5CH_3$ CH_3COOH $C_6H_5CO\text{-}O\text{-}COCH_3$

$CH_3CHOHCH_3$

Even necessary bonds can be placed on the line:

$\overset{\displaystyle R\text{-}C\text{-}OH}{\underset{\displaystyle O}{\|}}$ *is better as*   RCOOH

- Three centered dots indicate association of an unspecified type (e.g., hydrogen bonding, bond formation, or bond breaking).

$C\cdots Pt$ $F\cdots H\text{-}NH_3$

- The mass number of an isotope is the left superscript to an atomic symbol. For isotopic labeling and isotopic substitution, several recommendations are applicable:

1. For isotopic labeling, the number and symbol may be enclosed in square brackets closed up to the compound name. For isotopic substitution, the number and symbol may be enclosed in parentheses closed up to the compound name. (These are IUPAC/IUB recommendations.)

$^{32}PO_4^{3-}$ $H_2N^{14}CONH_2$ (^{14}C)glucose

[^{15}N]alanine   [2-^{14}C]leucine   [2,8-^3H]inosine

(^{15}N)ammonia

The left superscript should not be used in an abbreviation.

[^{32}P]CMP   *not*   CM^{32}P

2. Isotopically labeled compounds may also be described by inserting the symbol in brackets into the name of the compound. (This is an IUPAC rule for inorganic nomenclature.)

hydrogen chloride[^{36}Cl]   sulfuric[^{35}S] acid[^2H]

3. The Boughton system does not distinguish between labeling and substitution. The isotopic variation is shown by placing the symbol for the isotope (with a subscript numeral to indicate the number of isotopic atoms) after the name or relevant portion of the name; locants are cited if necessary.

ethane-1-*d*-2-*t*   methane-*d*$_4$ acetamide-*1*-^{13}C-^{15}N

benzeneacetic-*carboxy*,α-$^{14}C_2$ acid

4-(2-propenyl-3-^{13}C-oxy)benzoic acid

- In narrative text, the spelled-out element name followed by its mass number may be used for isotopes. A hyphen is used between the name and number.

uranium-235   carbon-14

- Square brackets are used to indicate concentration with an element symbol or formula, not with a spelled-out name.

[Ca]   *not*   [calcium]

[NaCl]   *not*   [sodium chloride]

- The prefixes "cis", "trans", "sym", "asym" may be used with a chemical formula, connected by a hyphen, and in italic face.

  cis-[PtCl$_2$(NH$_3$)$_2$]

Chemical Reactions

- Single-line chemical reactions are treated like mathematical equations; that is, they are displayed and numbered, if numbering is needed. The sequential numbering system used should integrate both chemical and mathematical equations.

$$Cr(CO)_4 + CO \xrightarrow{k} Cr(CO)_5 \qquad (1)$$

$$NH_3 + HCOOH \longrightarrow NH_2CHO + H_2O \qquad (2)$$

$$(CH_6H_5)_2P\text{-}P(C_6H_5)_2 \xrightarrow{h\nu} 2(C_6H_5)_2P\cdot \qquad (3)$$

$$Fe(CO)_5 + OCH_3^- \underset{k_{-1}}{\overset{k_1}{\rightleftharpoons}} Fe(CO)_4(CO_2CH_3)^- \qquad (4)$$

- When solid, liquid, gas, or aqueous is indicated, the appropriate abbreviations should be on the line and in parentheses.

$$Ag(s) + H^+(aq) + Cl^-(aq) \longrightarrow AgCl(s) + \tfrac{1}{2}H_2(g) \qquad (5)$$

- Chemical reactions that include structures with rings should be treated as illustrations. They are discussed in Chapter 3.

Computer Terms

- Use all capital letters for most computer languages.

  BASIC   COBOL   APL

  Exceptions: Assembler, Pascal

- Use initial capital letters for names of programs.

  Unifac   Uniquac   Ortep   Multan   Symphony   Wordstar
  MacWrite

- Use all capital letters for commands.

GOTO	LIST	LET	SCROLL
IF…THEN	RUN	CANCEL	DELETE
PRINT	GOSUB	HOME	
EXECUTE	END	EXIT	

Word Usage

- Stick to the original meaning of words; don't use a word to express a thought if such usage is the fourth or fifth definition in the dictionary or if such usage is primarily literary.

- Avoid slang or jargon.

- If you have already presented your results at a symposium or other meeting and are now writing the paper for publication in a book or journal, delete all references to the meeting or symposium such as "Good afternoon, ladies and gentlemen", "This morning we heard", "in this symposium", "at this meeting", "I am pleased to be here". Such phrases would be appropriate only if you were asked to provide an exact transcript of a speech.

- Correct use of the articles "a" and "an" follows pronunciation of the words or abbreviations they precede.

  an NMR spectrometer   a nuclear magnetic resonance spectrometer

  Use "a" before an aspirated "h"; use "an" before the vowel sound of *a, e, i, o,* and *u.*

a house	a history	*but*	an hour	an honor
a union	a U-^{14}C	*but*	an ultimate	

- Use the simple past tense for actions that occurred in the past.

McCeney reported	*not*	McCeney has reported
The values were measured previously	*not*	have been measured
Early investigations were reviewed	*not*	have been reviewed
We recently showed	*not*	have showed recently

- Use first person where appropriate. Do not use plural if singular is appropriate. That is, use "I" for one author and "we" for two or more authors. First person is perfectly acceptable where it helps keep your meaning clear:

    Poor
    In this author's previous publication...

    Better
    In my previous publication...

    Poor
    The authors recently demonstrated...

    Better
    We recently demonstrated...

    Poor
    Robinson reported xyz, but the present authors found...

    Better
    Robinson reported xyz, but we found...

    However, phrases like "we believe", "we feel", "we concluded", and "we can see" are unnecessary, as are personal opinions.

- Use the proper subordinating conjunctions. "While" and "since" have strong connotations of time. Do not use them where you mean "although", "because", or "whereas".

    Poor
    Since solvent reorganization is a potential contributor, the selection of data is very important.

    Better
    Because solvent reorganization is a potential contributor, the selection of data is very important.

    Poor
    While the reactions of the anion were solvent dependent, the corresponding reactions of the substituted derivatives were not.

    Better
    Although the reactions of the anion were solvent dependent, the corresponding reactions of the substituted derivatives were not.

- Use "respectively" to compare two sequences in the same sentence.

    The excitation and emission were measured at 360 and 440 nm, respectively.

- Convert nouns to their adjectival forms when they are being used as adjectives.

> an NMR spectroscopic study　　*not*　　an NMR spectroscope study
>
> mass spectrometric detection　　*not*　　mass spectrometry detection
>
> microanalytical technique　　*not*　　microanalysis technique
>
> transmission electron microscopic method　　*not*　　transmission electron microscope method

- Do not use "respectively" when you mean "separately".

> *Incorrect*
> The electrochemical oxidations of chromium and tungsten tricarbonyl complexes, respectively, were studied.
>
> *Correct*
> The electrochemical oxidations of chromium and tungsten tricarbonyl complexes were studied separately.

- Use "due to" only to modify a noun or pronoun.

> *Incorrect*
> Delays resulted due to equipment failure.
>
> *Correct*
> Delays due to equipment failure are unavoidable.
>
> *Also correct*
> The delays were due to equipment failure.

- Be sure that the antecedents of the pronouns "this" and "that" are clear. If there is a chance of ambiguity, use a noun to clarify your meaning.

> *Ambiguous*
> The photochemistry of transition-metal carbonyls has been the focus of many investigations. This is due to the central role that metal carbonyls play in various reactions.
>
> *Unambiguous*
> The photochemistry of transition-metal carbonyls has been the focus of many investigations. This interest is due to the central role that metal carbonyls play in various reactions.

- Use "less" with number and unit of measure combinations, because they are regarded as singular.

> less than 5 mg　　less than 3 days

- Avoid misuse of prepositional phrases introduced by "with".

  Poor
  Nine deaths from leukemia occurred, with six expected.

  Better
  Nine deaths from leukemia occurred, and six had been expected.

  Poor
  Of the 20 compounds tested, 12 gave positive reactions, with three being greater than 75%.

  Better
  Of the 20 compounds tested, 12 gave positive reactions, and three of these were greater than 75%.

  Poor
  Two weeks later, six more animals died, with the total rising to 25.

  Better
  Two weeks later, six more animals died, and the total was then 25.

- Use the more accurate terms "greater than" or "more than" rather than the imprecise "over" or "in excess of ".

greater than 50%	*not*	in excess of 50%		
more than 100 samples	*not*	over 100 samples		
more than 25 mg	*not*	in excess of	*or*	over 25 mg

- Do not use a solidus to mean "and" and "or".

  Incorrect
  Hot/cold extremes will damage the samples.

  Correct
  Hot and cold extremes will damage the samples.

  "And/or" can usually be replaced by either "and" or "or".

- Use split infinitives to avoid awkwardness or ambiguity.

  Awkward
  ...to assist financially the student who is considering a career in chemistry.

  Better
  ...to financially assist the student who is considering a career in chemistry.

- Use "whether" to introduce at least two alternatives; use "whether or not" to mean "regardless of whether".

    Incorrect
    I don't know whether or not to repeat the experiment.

    Correct
    I don't know if I should repeat the experiment.

    Also correct
    The experiment must be repeated whether the results are positive or negative.

    Also correct
    Whether or not the results are positive, the experiment must be repeated.

- Use "comprise" to mean "contain"; the whole comprises the parts (not the whole is comprised of the parts).

    Incorrect
    A book is comprised of chapters.

    Correct
    A book comprises chapters.

    Incorrect
    Our research was comprised of three stages.

    Correct
    Our research comprised three stages.

- Use "fewer" and "less" correctly: "fewer" refers to number; "less" refers to quantity.

    fewer than 50 animals   fewer than 100 samples

    less product   less time   less work

Nonsexist Language

The U.S. Government and many publishers have gone to great effort to discourage the use of sexist language in their publications. Sexist language is also a concern of many chemists. Recent style guides and writing guides urge editors and writers to choose terms that do not reinforce outdated sex roles. Nonsexist language can be accurate and unbiased, and not necessarily awkward.

The most common problems are with the noun "man" and the pronouns "he" and "his", but there are usually several nonsexist alternatives for these words. Choose an alternative carefully and keep it consistent with the context.

- Instead of "man", use "people", "humans", "human beings", or "human species", depending on your meaning.

    Sexist
    The effects of compounds of I–X were studied in rats and man.

    Nonsexist
    The effects of compounds I–X were studied in rats and humans.

    Sexist
    Men working in hazardous environments are often unaware of their rights and responsibilities.

    Nonsexist
    People working in hazardous environments are often unaware of their rights and responsibilities.

    Sexist
    Man's search for beauty and truth has resulted in some of his greatest accomplishments.

    Nonsexist
    The search for beauty and truth has resulted in some of our greatest accomplishments.

- Instead of "manpower", use "workers", "staff", "work force", or "personnel", depending on your meaning.

- Instead of "he" and "his", use plural ("they" and "theirs") or first person ("we", "us", and "ours"). You can delete "his" and replace it with "the" or nothing at all. "His or her", if not overused, is not terribly unpleasant. You can also recast the sentence.

    Sexist
    The principal investigator should place an asterisk after his name.

    Nonsexist
    Principal investigators should place asterisks after their names.

    Nonsexist
    If you are the principal investigator, place an asterisk after your name.

    Nonsexist
    The name of the principal investigator should be followed by an asterisk.

• Instead of "wife", use "family" or "spouse" where appropriate.

> *Sexist*
> The work of professionals like chemists and doctors is often so time-consuming that their wives are neglected.
>
> *Nonsexist*
> The work of professionals like chemists and doctors is often so time-consuming that their families are neglected.
>
> *Sexist*
> The society member and his wife...
>
> *Nonsexist*
> The society member and spouse...

Wordiness

• Be brief. Wordiness usually adds nothing but confusion, and the resulting paper is very expensive to typeset and to print.

• Omit phrases such as "it was found that", "it was demonstrated that", and "it would appear that".

• Avoid unnecessary words:

the results *tend to* suggest	they are *both* alike
estimated at *about* 10%	throughout the *entire* experiment
such as copper, iron, *etc.*	two *equal* halves
bright red *in color*	*very* unique
fewer *in number*	*as* yet
oval *in shape*	*as to* whether

• Wordy expressions may often be replaced by a simple word:

Wordy	*Better*
owing to the fact that	because
subsequent to	after
on the order of	about
in the near future	soon
at the present time	now
by means of	by
it appears that	apparently
of great importance	important
in consequence of this fact	therefore
a very limited number of	few
in spite of the fact that	although

References

Citing in Text

- References may be cited in text in two ways: by number or by author name and date.

> Oscillation in the reaction of benzaldehyde with oxygen was reported previously.[3]
>
> Oscillation in the reaction of benzaldehyde with oxygen was reported previously (3).
>
> The primary structure of this enzyme has also been determined (Dardel et al., 1984).

- References are cited by superscript numbers without parentheses and without spaces in *Accounts of Chemical Research*, *Chemical Reviews*, *Inorganic Chemistry*, *Journal of the American Chemical Society*, *Journal of Chemical Information and Computer Sciences*, *Journal of Medicinal Chemistry*, *The Journal of Organic Chemistry*, *The Journal of Physical Chemistry*, *Langmuir*, *Macromolecules*, and *Organometallics*.

- References are cited by numbers on the line, underlined (for italic), in parentheses, and with spaces in ACS books, *Analytical Chemistry*, *Environmental Science & Technology*, and *Journal of Chemical and Engineering Data*.

- References are cited by author name in *Biochemistry*, *Industrial & Engineering Chemistry Fundamentals*, *Industrial & Engineering Chemistry Process Design and Development*, *Industrial & Engineering Chemistry Product Research and Development*, and *Journal of Agricultural and Food Chemistry*.

- With numerical reference citations, start with 1 and number consecutively throughout the paper. If a reference is repeated, do not give it a new number; use the original reference number. Use only numbers, not combinations of numbers and letters (1 and 2, not 1a and 1b).

- When citing more than one reference at one place, list the numbers in ascending order and separate them by commas (without spaces as superscripts; with spaces on line), or if they are part of a consecutive series, use a dash for three or more.

> ...in the literature[2,5,8]   were reported[3-5,10]
>
> ...in the literature (2, 5, 8)   were reported (3-5, 10)

- Even when references are cited by number, you may also use an author name, directly followed by the reference number.

  Jensen (3) reported oscillation in the reaction of benzaldehyde with oxygen.

- In both systems, if a reference has two authors, give both names (Allison and Perez[12]); if a reference has more than two authors, give only the first name listed and "et al." [Johnson et al. (*12*)].

- If citing more than one reference by the same principal author and various coauthors, use the principal author's name followed by "and co-workers" (Brown and co-workers[10,11]). If it is necessary to distinguish among several references by the same authors published in the same year, add a, b, c to the year (Jones and Smith, 1980c; Steele et al., 1986b).

- Cite the reference in a logical place in the sentence.

recent investigations (cite)	recently were demonstrated (cite)
other developments (cite)	a molecular mechanics study (cite)
was reported recently (cite)	Marshall and Levitt's approach (cite)
as described previously (cite)	the procedure of Riesberg et al. (cite)
previous results (cite)	

Listing

- Collate all references at the end of the manuscript in numerical order if cited by number and in alphabetical order if cited by author.

- The author is responsible for the accuracy and completeness of all references. The author should check all parts of each reference listing against the original document.

- A reference must include certain minimum data: for journals—author, abbreviated journal title, year of publication, volume number, and initial page of cited article (the complete span is better); for books—author or editor, book title, publisher, city of publication, and year of publication.

- For material other than books and journals, enough information must be provided so that the source can be identified and located.

- In page spans, use all digits.

2–15	12–19	44–49	103–107	108–117	234–236
345–359	1376–1382	2022–2134			

Journals

Chemical Abstracts Service Source Index (*CASSI*) and its quarterly supplements provide a complete list of recommended journal abbreviations. The following is a short list of the most commonly cited journals as they appear in 1985. Note that one-word titles are not abbreviated (e.g., *Biochemistry, Nature, Science*).

Acc. Chem. Res.	*Chem. Abstr.*
Acta Crystallogr., Sect. A	*Chem. Ber.*
Acta Crystallogr., Sect. B	*Chem. Biol. Interact.*
Acta Crystallogr., Sect. C	*Chem. Eng. News*
AIChE J.	*Chem. Eng. (N.Y.)*
AIChE Symp. Ser.	*Chem.-Ing.-Tech.*
Anal. Biochem.	*Chem. Lett.*
Anal. Chem.	*Chem. Listy*
Anal. Chim. Acta	*Chemotherapy (Tokyo)*
Anal. Lett.	*Chem. Pharm. Bull.*
Angew. Chem.	*Chem. Phys.*
Ann. Chim. (Rome)	*Chem. Phys. Lett.*
Antimicrob. Agents Chemother.	*Chem. Rev.*
Appl. Opt.	*CHEMTECH*
Appl. Phys. Lett.	*Chem.-Ztg.*
Appl. Spectrosc.	*Clin. Chem. (Winston-Salem, N.C.)*
Arch. Biochem. Biophys.	*Cold Spring Harbor Symp. Quant. Biol.*
Aust. J. Chem.	*Collect. Czech. Chem. Commun.*
Ber. Bunsenges. Phys. Chem.	*C. R. Hebd. Seances Acad. Sci.*
Biochem. Biophys. Res. Commun.	*Dokl. Akad. Nauk SSSR*
Biochemistry	*Drug Metab. Dispos.*
Biochem. J.	*Electrochim. Acta*
Biochem. Pharmacol.	*Endrocrinology (Baltimore)*
Biochim. Biophys. Acta	*Environ. Sci. Technol.*
Biochimie	*Eur. J. Biochem.*
Biofizika	*Exp. Cell Res.*
Biokhimiya	*Experientia*
Biopolymers	*FEBS Lett.*
Bull. Chem. Soc. Jpn.	*Fed. Proc., Fed. Am. Soc. Exp. Biol.*
Bull. Soc. Chim. Belg.	*Gazz. Chim. Ital.*
Bull. Soc. Chim. Fr.	*Helv. Chim. Acta*
Cancer Res.	*Hoppe–Seyler's Z. Physiol. Chem.*
Can. J. Biochem.	*IEEE J. Quantum Electron.*
Can. J. Chem.	*Ind. Eng. Chem. Fundam.*
Carbohydr. Res.	*Ind. Eng. Chem. Process Des. Dev.*

Ind. Eng. Chem. Prod. Res. Dev.
Inorg. Chem.
Inorg. Chim. Acta
Int. J. Chem. Kinet.
Int. J. Mass Spectrom. Ion Phys.
Int. J. Quantum Chem.
Isr. J. Chem.
Izv. Akad. Nauk SSSR, Ser. Khim.
J. Agric. Food Chem.
J. Am. Chem. Soc.
J. Assoc. Off. Anal. Chem.
J. Bacteriol.
J. Biochem. (Tokyo)
J. Biol. Chem.
J. Chem. Eng. Data
J. Chem. Inf. Comput. Sci.
J. Chem. Soc., Chem. Commun.
J. Chem. Soc., Dalton Trans.
J. Chem. Soc., Faraday Trans. 1
J. Chem. Soc., Faraday Trans. 2
J. Chem. Soc., Perkin Trans. 1
J. Chem. Soc., Perkin Trans. 2
J. Chim. Phys. Phys.-Chim. Biol.
J. Chromatogr.
J. Electroanal. Chem.
J. Electrochem. Soc.
J. Endocrinol.
J. Fluorine Chem.
J. Heterocycl. Chem.
J. Lipid Res.
J. Macromol. Sci., Chem.
J. Magn. Reson.
J. Med. Chem.
J. Mol. Biol.
J. Mol. Spectrosc.
J. Organomet. Chem.
J. Org. Chem.
J. Pharmacol. Exp. Ther.
J. Pharm. Pharmacol.
J. Pharm. Sci.
J. Phys. A: Math Gen.
J. Phys. B
J. Phys. C

J. Phys. Chem.
J. Phys. Chem. Ref. Data
J. Phys. Chem. Solids
J. Physiol. (London)
J. Phys. (Orsay, Fr.)
J. Polym. Sci., Polym. Chem. Ed.
J. Polym. Sci., Polym. Lett. Ed.
J. Polym. Sci., Polym. Phys. Ed.
J. Polym. Sci., Polym. Symp.
Langmuir
Liebigs Ann. Chem.
Lipids
Macromolecules
Makromol. Chem.
Mol. Pharmacol.
Monatsh. Chem.
Nature (London)
Naturwissenschaften
Organometallics
Org. Mass Spectrom.
Phys. Lett. A
Phys. Rev. A
Phys. Rev. Lett.
Polym. J.
Polym. Prepr., Am. Chem. Soc. Div.
   Polym. Chem.
Prepr. Pap.—Am. Chem. Soc., Div. Fuel
   Chem.
Proc. Natl. Acad. Sci. U.S.A.
Proc. Soc. Exp. Biol. Med.
Pure Appl. Chem.
Science (Washington, D.C.)
Spectrochim. Acta, Part A
Spectrosc. Lett.
Steroids
Synth. Commun.
Tetrahedron
Tetrahedron Lett.
Theor. Chim. Acta
Usp. Khim.
Z. Anorg. Allg. Chem.
Zh. Fiz. Khim.
Zh. Neorg. Khim.

Zh. Obshch. Khim.	*Z. Naturforsch., B: Anorg. Chem., Org. Chem.*
Zh. Org. Khim.	*Z. Phys. Chem. (Leipzig)*
Z. Naturforsch., A	*Z. Phys. Chem. (Wiesbaden)*

- In the following examples, the journal titles and volume numbers appear in italics, so they should be underlined in the manuscript. The years are set in boldface type, so they should be underscored with a wavy line.

  Fletcher, T. R.; Rosenfeld, R. N. *J. Am. Chem. Soc.* **1985**, *107*, 2203–2212.

  Huffman, J. C.; Lewis, L. N.; Caulton, K. G. *Inorg. Chem.* **1980**, *19*, 2755.

  Carson, M. A.; Atkinson, K. D.; Waechter, C. J. *J. Biol. Chem.* **1982**, *257*, 8115–8121.

  Rose, L. M.; Hyka, J. *Ind. Eng. Chem. Process Des. Dev.* **1984**, *23*, 429–437.

- In references to journals that begin every issue with page 1, include the issue number in parentheses following the volume number.

  Stinson, S. C. *Chem. Eng. News* **1985**, *63*(25), 26.

- Indicate when reference is made to an abstract of an article. If possible, give both the original article and the abstract.

  Roe, D. D. *Zh. Fiz. Khim.* **1985**, *72*, 1234; *Chem. Abstr.* **1985**, *78*, 122a.

- Indicate when a reference is to the English translation of an article printed in a non-English-language journal. If possible, also include reference to the original article.

  Doe, A. B. *J. Gen. Chem. USSR (Engl. Transl.)* **1985**, *55*, 2050; *Zh. Obshch. Khim.* **1985**, *55*, 2100.

Books

- In publishers' names, delete words such as "Company", "Inc.", "Publisher", and "Press".

- In the following examples, book titles appear in italics, so they should be underlined in the manuscript.

• Use the following abbreviations and spelled-out forms as indicated, and follow the capitalization shown.

>Abstract (spell out)
>Chapter (spell out)
>ed. (for edition)
>Ed., Eds. (for editor, editors)
>No. (for number)
>p, pp (for page, pages, with no periods)
>paper (spell out)
>Part (spell out)
>Vol. (for a specific volume, Vol. 4)
>vols. (for number of volumes, 4 vols.)

BOOKS WITHOUT EDITORS

>Chum, H. L.; Baizer, M. M. *The Electrochemistry of Biomass and Derived Materials*; ACS Monograph 183; American Chemical Society: Washington, DC, 1985; pp 134–157.

>Stothers, J. B. *Carbon-13 NMR Spectroscopy*; Academic: New York, 1972; Chapter 2.

>Bockris, J. O.; Reddy, A. K. N. *Modern Electrochemistry*; Plenum: New York, 1970; Vol. 2, p 132.

>Littman, M.; Yeomans, D. K. *Comet Halley: Once in a Lifetime*; American Chemical Society: Washington, DC, 1985; p 23.

BOOKS WITH EDITORS

>*Mapping Strategies in Chemical Oceanography*; Zirino, A., Ed.; Advances in Chemistry 209; American Chemical Society: Washington, DC, 1985.

>Geacintov, N. E. In *Polycyclic Hydrocarbons and Carcinogenesis*; Harvey, R. G., Ed.; ACS Symposium Series 283; American Chemical Society: Washington, DC, 1985; pp 12–45.

>Kolar, G. F. In *Chemical Carcinogens*, 2nd ed.; Searle, C. E., Ed.; ACS Monograph 182; American Chemical Society: Washington, DC, 1984; Vol. 2, Chapter 14.

>Golay, M. J. E. In *Gas Chromatography*; Desty, D. H., Ed.; Butterworths: London, 1958; p 36.

>Jennings, K. R. In *Mass Spectroscopy*; Johnstone, R. A. W., Senior Reporter; Specialist Periodical Report; The Chemical Society: London, 1977; Vol. 4, Chapter 9.

Theses

- Give the person's name, the level of thesis, the university, and the date as specifically as possible.

    Kanter, H. Ph.D. Thesis, University of California at San Francisco, Dec. 1984.

    Fleissner, W. Ph.D. Dissertation, University of Tennessee, 1984.

Patents

- Give the person's name, the patent number, and the year. If possible, include the *Chemical Abstracts* reference as well.

    Norman, L. O. U.S. Patent 4 379 752, 1983.

    Jordan, O. D. Br. Patent 2 081 298, 1982.

    Lyle, F. R. U.S. Patent 5 973 257, 1985; *Chem. Abstr.* **1985**, 65, 2870.

Government Publications

Government publications can be pamphlets, brochures, books, maps, journals, and almost anything else that can be printed. They may have authors or editors, who may be individuals or offices, or they may be unauthored. They are published by the specific agencies, but they are usually (not always) available through the Government Printing Office rather than the agency. To enable others to find the publication, the American Library Association suggests that you include as much information as you have. The following are examples of the most commonly cited types. Book and journal titles are set in italic type; titles of other publications and laws are set in Roman face and in quotes.

    Energy Alternatives and the Environment: 1979; U.S. Environmental Protection Agency. Office of Research and Development. Office of Environmental Engineering and Technology. U.S. Government Printing Office: Washington, DC, 1979; EPA-600/9-80-009.

    Sherma, J.; Beroza, M. *Manual of Analytical Quality Control for Pesticides and Related Compounds*; U.S. Environmental Protection Agency. U.S. Government Printing Office: Washington, DC, 1979; EPA-600/1-79/008.

    Interdepartmental Task Force on PCBs. *PCBs and the Environment*; U.S. Government Printing Office: Washington, DC, 1972; COM 72.10419.

    "Toxic Substances Control Act"; Public Law 94-469, 1976.

    Pesticides Analytical Manual; U.S. Department of Health, Education, and Welfare. Food and Drug Administration. U.S. Government Printing Office: Washington, DC, 1982, Vol. 1.

National Handbook of Recommended Methods for Water Data Acquisition; Office of Water Data Coordination, U.S. Geological Survey: Reston, VA, 1977; Chapter 5.

Reactor Safety Study: An Assessment of Accident Risks in U.S. Commercial Power Plants; NUREG 75/014; Nuclear Regulatory Commission: Washington, DC, 1975.

Fed. Regist. **1984**, *50*, 1984.

Diet and Dental Health: A Study of Relationships; U.S. National Center for Health Statistics; DHHS Pub. No. PHS 82-1675, U.S. Government Printing Office: Washington, DC, 1982.

"The Measurement of the Catalytic (Activity) Concentration of Seven Enzymes in NBS Human Serum SRM 909"; *NBS Spec. Publ. (U.S.)* **1983**, No. 260-88.

State and local governments and their agencies also publish.

Tofflemire, T. J.; Quinn, S. O. "PCBs in the Upper Hudson River"; Technical Report No. 56; New York State Department of Environmental Conservation: Albany, NY, 1977.

"Air Quality Aspects of the Development of Offshore Oil and Gas Resources"; California Air Resources Board: Sacramento, CA, 1982.

Reports

Schneider, A. B. Technical Report No. 1234-56, 1985; ABC Company, New York.

Morgan, M. G. "Technological Uncertainty in Policy Analysis"; final report to the National Science Foundation on Grant PRA-7913070; Carnegie-Mellon University: Pittsburgh, PA, 1982.

Abstracts of Meeting Papers

Baisden, P. A. *Abstracts of Papers*, 188th National Meeting of the American Chemical Society, Philadelphia, PA; American Chemical Society: Washington, DC, 1984; NUCL 9.

Goodman, P. W. *Abstracts of Papers*, International Chemical Congress of Pacific Basin Societies, Honolulu, HI; American Chemical Society: Washington, DC, 1984; Abstract 05F14.

Material Presented Orally

Ford, W. T. Presented at the 189th National Meeting of the American Chemical Society, Miami, FL, April 1985; paper ORGN 79.

Rose, J. J. Presented at the Pittsburgh Conference, Atlantic City, NJ, March 1983; paper 707.

Wilkins, C. L. Presented at the Pacific Conference on Chemistry and Spectroscopy, Pasadena, CA, Oct. 1983.

Castro, M. E.; Russell, D. H. Presented at the 32nd Annual Conference on Mass Spectrometry and Allied Topics, San Antonio, TX, 1984.

Unpublished Materials

Material accepted for publication but not yet published belongs in lists of literature cited. Strictly speaking, material submitted for publication but not yet accepted and personal communications are not part of the literature and therefore do not belong in lists labeled "Literature Cited", but may be included in footnotes that also include other notes. In ACS books and *Biochemistry*, material submitted but not accepted and personal communications are parenthetical notes in text, not numbered and not included in the "Literature Cited".

● Material accepted for publication but not yet published:

Roe, A. B. *J. Pharm. Sci.*, in press.

● Material submitted but not yet accepted:

Roe, A. B., submitted for publication in *J. Pharm. Sci.*

● Personal communications:

Doe, C. D., The State University of New York at Buffalo, personal communication, 1985.

Doe, C. D., The State University of New York at Buffalo, unpublished results.

3
Chapter

Illustrations and Tables

This chapter is intended to be a general guide to using and providing art and tables for a scientific paper. Certain style points and other requirements differ from journal to journal or publisher to publisher. For ACS publications, it is important to consult recent issues of journals as well as the Guide, Notes, or Instructions for Authors that appear in each journal's first issue of the year, or the Requirements for Manuscripts for books.

Illustrations

When To Use Illustrations

Illustrations can play a major role in highlighting and clarifying results and data. Appropriate, well-drawn illustrations can substantially increase comprehension of the text and can convey trends, comparisons, and relationships more clearly than text. However, illustrations that do not clarify the discussion but merely repeat data already presented in text, and illustrations that are poorly rendered not only decrease comprehension, but also cause confusion. Illustrations are time-consuming and costly for authors to prepare and also for publishers to reproduce, so they should be worth the time, effort, and money.

How To Draw Line Artwork

Line art usually consists of only black markings on a white background and no shades of gray.

It is probably best to have a professional graphic artist or draftsman prepare your artwork. If you do it yourself, you will need some art supplies. Use white vellum paper or Bristol board. Do not use ordinary graph paper, but you may use graph paper ruled in "nonphoto" or "nonrepro" blue. Nonphoto blue ink will not reproduce in printing. Do not use tracing paper or Mylar. Use only black ink on white paper, and not other colors of ink or paper.

To draw the lines, use black ink and a rapidograph pen (No. 0) or a Leroy lettering pen (No. 1). For the axes, you will need a T-square or triangle. For the curves, you will need a French curve. Because thin lines will break in reproduction, you should draw all lines to a medium line weight.   Use a slightly heavier line for the curves than for the axes.

For lettering, use dry transfer type (such as Letraset, Zipatone, or Chartpak) or a mechanical lettering set (such as Leroy or Wrico). Press type is neat and easy to use, but if it is not applied properly it can flake off. Mechanical lettering is inked, so it will not flake off, but it can look uneven if it is not inked properly.

For symbols, use dry transfer type (such as Copyaid) or black ink and a template. Use simple geometric symbols such as circles, squares, diamonds, and triangles. Do not use unusual shapes or freehand marks. Use only all-clear and all-dark symbols; do not use half-dark symbols or symbols with intersecting lines. These two kinds are often not easily distinguished from their clear and dark counterparts. Use symbols instead of lines with dashes of different lengths. Symbols are usually much clearer and easier to follow. If you cannot obtain different kinds of symbols, a good alternative is to label the curves with letters and then refer to them in the caption as "curve A", "curve B", etc.

Shading in line art can be accomplished with diagonal or crosshatched lines. Commercial products in a variety of patterns and tones can also be used to create shading in line art.

You must also keep the following considerations in mind:

Usually, it is best to draw artwork larger than the final size required for the journal or book. Then, in production, the artwork is reduced to fit. When artwork is reduced, minor imperfections disappear, and the lines are stronger. This system works best when the artwork is 50% larger than the desired final size. Furthermore, artwork that is 50% larger is still small enough to be mailed and handled easily.

One point that is obvious after it is said, but sometimes not before, is that when the art gets reduced, everything gets proportionately smaller: the length, width, type, symbols, and lines. Consequently, all of these elements should be in proportion to each other.

The overall size, the type size, and the symbol size should all be 50% larger. (See page 125.) The symbols should be the size of a lower-case letter "oh". If the type is very large and the symbols are very small, when the art is reduced to fit, the type will be an appropriate size, but the symbols will be so small that they will be unintelligible. If the symbols are large and the type is small, reduction will render the type illegible. Likewise, if the overall size is very large and the type and symbols are small, reduction to fit the column will render the type and symbols unreadable. Table IX demonstrates the relationships between components of artwork as drawn and as reduced for publication. Figures 1-5 demonstrate artwork at the proper size before and after reduction and poorly drawn artwork.

Table IX. Column Widths and Type Sizes of Illustrations in ACS Publications, As Published and 50% Larger

Publication	Column Width			Type Size, points[a]	Symbol Size, mm
	picas	*inches*	*cm*		
Books					
as published	27	4.5	11	9	2
50% larger	41	6.75	17	14	3
Journals and magazines, two-column format single column					
as published	20	3.25	8.25	8	2
50% larger	30	5	13	12	3
double column					
as published	41	7	17	8	2
50% larger	62	10.5	26	12	3
Magazines, three-column format single column					
as published	13	2	5	8	2
50% larger	19.5	3.25	8.25	12	3
double column					
as published	27.5	4.5	11	8	2
50% larger	41	6.75	17	12	3
triple column					
as published	42	7	17.5	8	2
50% larger	64	10.5	27	12	3

NOTE: The maximum usable column lengths are 6.5 inches (16.5 cm) in books and 9.5 inches (24 cm) in journals and magazines.

In production, artwork will be reduced to appropriate sizes, often less than the full column width.

[a] Subscripts and superscripts are usually 2 points smaller.

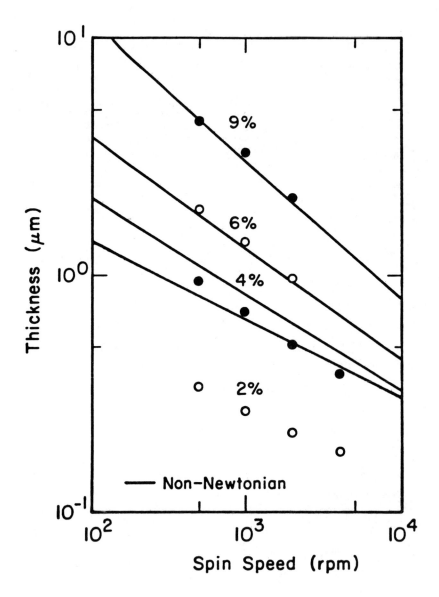

Figure 1A. Artwork at the proper original size.

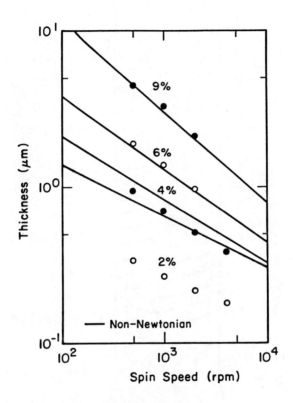

Figure 1B. Figure 1A as reduced for publication.

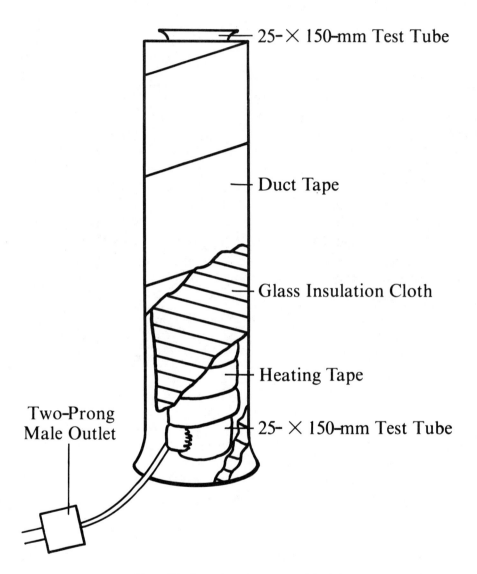

Figure 2A. Artwork at the proper original size.

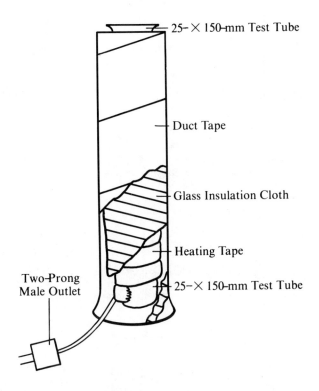

Figure 2B. Figure 2A as reduced for publication.

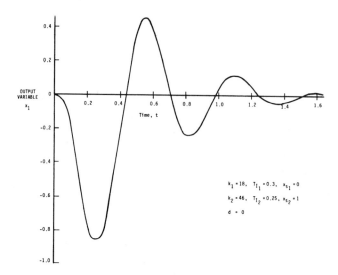

Figure 3A. The type in this artwork is too small in relation to the other elements.

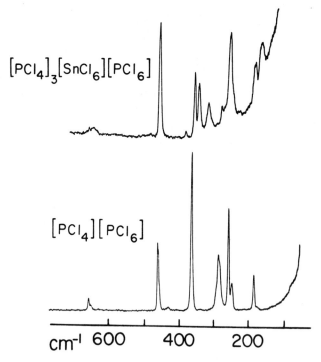

Figure 3B. The type in this artwork is too large in relation to the other elements.

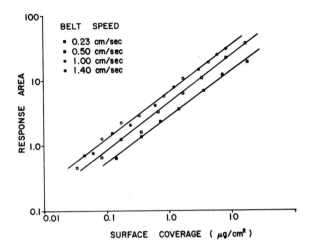

Figure 4A. The symbols in this artwork are too small in relation to the other elements.

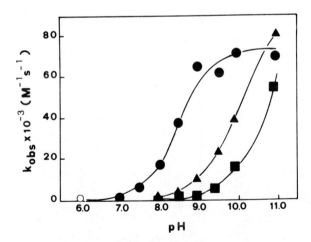

Figure 4B. The symbols in this artwork are too large in relation to the other elements.

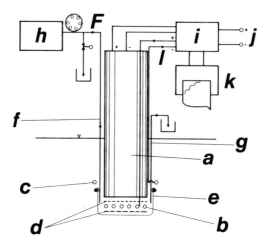

Figure 5A. The lines in this artwork are too thin in relation to the other elements. The type is also too large.

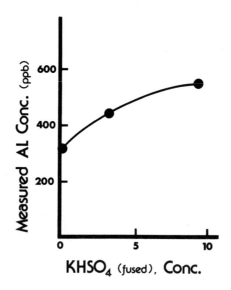

Figure 5B. The lines in this artwork are too thick in relation to the other elements.

Type Size and Typeface

In publishing, type is measured in *points*; space is measured in *picas*. There are 72 points to an inch; there are 6 picas to an inch. The size of the type you are reading is 10 points. The column is 26 picas wide, and the text page is 42 picas long.

You can buy a pica ruler for measuring space, but there is no ruler for measuring type: you must compare it to known type sizes. When you buy type, you must specify the type size you want, in points. Mechanical lettering kits come with several sizes of templates, from 6 to 50 points, which are labeled 60CL to 500CL, respectively. You may not need to buy the whole kit, because all of the type on your art should be the same size, except for subscripts and superscripts. You probably will need only a few sizes of type. For example, for all ACS books and journals, you would need 12- and 14-point type for the words and 10- and 12-point type for the subscripts and superscripts.

You must also specify the *typeface*, which is the style or design of the letters. There are literally hundreds of typefaces, but standard faces like Times Roman, Century, or Helvetica are best for scientific art.

Type also comes in different weights. Most of the type you are reading is lightface, **and this is boldface type**.

Finally, type may be *italic* or Roman.

Generally, you should use lightface, Roman type.

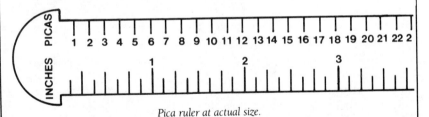

Pica ruler at actual size.

This is 12-point Times Roman. $^{14}C_6H_6$ shows subscripts and superscripts.

This is 12-point Helvetica. $^{14}C_6H_6$ shows subscripts and superscripts.

This is 14-point Times Roman. $^{14}C_6H_6$ shows subscripts and superscripts.

This is 14-point Helvetica. $^{14}C_6H_6$ shows subscripts and superscripts.

If you cannot follow the guidelines in Table IX exactly, then, at the very least, keep all art for the same manuscript consistent in type size and face, symbol size, and line weight.

Combining curves that have the same set of axes saves time, space, and money, but do not put more than four or five curves in one set of axes. Leave sufficient space between curves; they should not overlap so much that the symbols are indistinguishable.

Keep the figure compact; have axes only long enough to define the curve. For example, if the highest point on the curve is 14, then 15 should be the highest axis label. Furthermore, the origin or lowest point on the axes does not have to be zero. For example, if the lowest point on the curve is 4, then 3 should be the lowest axis label. Put grid marks on the axes to indicate the positions of the numbers.

Label the axes with the parameter or variable being measured, the units of measure, and the scale. Use initial capital letters only, not all capitals. Place the units of measure in parentheses. Place all labels outside the axes and parallel to them. Letters and numbers should read from left to right and from bottom to top. Do not place arrowheads on the ends of the axis lines; simple straight lines are adequate. Use only two axes, one horizontal and one vertical; do not draw the top horizontal and right vertical lines to make a box.

Keep the illustration clear and simple. Keep wording to a minimum.

Leave clear margins of at least 1 inch (2.5 cm) on all sides of the artwork. The editors need the space for marking codes and identification and for handling.

If possible, have your original line art statted. A *stat* (also called Photostat or PMT) is a photographic copy on a special matte-finish paper. By submitting stats, you can avoid losing type that may flake off the original, and corrections made with opaquing fluid will not reproduce.

Usually, computer-generated figures are not acceptable because the lines are so light that they are lost in reproduction, the size of the type is too small, and the overall size is too large. Drawings produced on older dot matrix printers are usually unacceptable for these reasons. However, if these elements can be corrected, computer-generated art can be used. The quality of computer-generated art depends on the plotter, pens, paper, and the graphics program used to create the pictures. Hardware and software are being improved every day. If your printer uses a pen with a sufficiently thick point and a high-quality glossy paper, and if you can follow the size guidelines in Table IX, then you can get a good image that will be acceptable.

Continuous-Tone Prints

All black and white photographs contain shades of gray and are called *continuous-tone prints*. They must be reproduced by a method that converts shades of gray into dot patterns suitable for platemaking. The various shades of gray are achieved with dot patterns (screens) of various densities: the greater the density, the darker the gray. If you look at a newspaper photograph, you can see the dots easily. However, you would need a magnifying glass to see the dots in a book or journal photograph because book and journal printers use finer screens. Nevertheless, there are dots in any photograph that has been printed. A printed photograph is called a *halftone*, and halftones cannot be rescreened. Therefore, you must submit only original black and white photographs. Furthermore, slides and photocopies of photographs also cannot be screened and are thus not usable for reproduction. If you have slides, you must have them converted to black and white prints.

Photographs always lose some detail in reproduction, so it is better to submit photographs that have slightly more contrast and are somewhat larger than the printed photograph will be. More contrast can be achieved by placing the object against a light or dark background. Also, most commercial photography laboratories can adjust contrast. When you get your film developed, you should tell them that the photograph will be used in a publication.

Most flaws on a continuous-tone print will be reproduced in the printed photograph. A continuous-tone print can be marred by indentations left by a paper clip or by writing on the back with a sharp-pointed pen. Obviously, folding and stapling will mar the print. Printers cannot correct such flaws, nor can they selectively omit them. Therefore, be extremely careful with continuous-tone prints because they cannot be fixed, and you may have to reshoot the photograph.

If you have a photograph of a person, you must have written permission from the person to reproduce the photograph. Likewise, if you have a photograph of commercial equipment, you must have written permission from the manufacturer to reproduce the photograph.

Because the continuous-tone process is a method for reproducing shades of gray, dark lines and type are often weakened in continuous-tone photographs. Therefore, stats of line art are usually better, except when the lines are not dark enough. In that case, a continuous-tone photograph can enhance the contrast (for example, computer-generated art).

In some cases, color is necessary to comprehend the material or the results. Color photographs must be "color separated", and four separate plates must be made in order to print in color. Because of this extra work, color printing is extremely expensive. Most publications cannot afford to incur these extra costs; authors must pay for color. Procedures differ for different publications; you must check with the specific editorial office.

How To Submit Illustrations

For line art, you should submit the original camera-ready artwork, stats, or glossy photographs. For continuous-tone prints, you must submit the original photographs. Submit all original art with the original manuscript, and photocopies with each copy of the manuscript. Place each figure on a separate sheet; do not incorporate figures into the text, but group them at the end of the paper.

Mark the figure number and author's name either on the back of the artwork or on the front, about an inch (2.5 cm) clear of the image area. Mark the top of the illustration with the word "top" when it would not be obvious which side goes up. Do not fold or roll artwork or photographs; protect them with cardboard or heavy paper when you mail them. Do not staple, clip, or punch holes in photographs or any artwork.

When marking the back of a photograph, write very lightly and use a soft pencil. The pressure of a ballpoint pen will indent the photo, the writing will show on the photograph face, and it will be reproduced in processing.

You could also identify photographs by marking a pressure-sensitive label and then affixing it to the back of the photograph.

How To Number and Cite Illustrations

Number figures sequentially with Arabic numerals, in order of discussion in the text. Every figure must be discussed in the text; refer to them as "Figure 1", "Figure 2", etc. Chemical structures and schemes should not be numbered as figures; they should be numbered separately.

Figure Captions

Every figure must have a caption that includes the figure number and a brief description, preferably one or two sentences or descriptive phrases. The caption should be understandable without reference to the text and should not include new material that is not in the text. Use similar wording for captions of related figures. Generally, put the keys to symbols in the caption, not in the artwork, where they tend to give a cluttered appearance. Make sure that the symbols and abbreviations in the caption agree with those in the figure itself and in the text. If the artwork contains unusual symbols, and these symbols may not be available to the typesetter, then you should identify the symbols within the artwork.

Submit the figure captions separately from the artwork, typed double spaced, all on a page together. If you place the caption on the art, the editors usually will delete it and typeset the caption according to the publication's style.

Credit lines for art reproduced from previously published work must appear at the end of the caption: "(Reproduced with permission from reference 10. Copyright 1985 Publisher.)"

Chemical Structures and Schemes

When To Use Structures and Schemes

Structures and reaction schemes are essential in some papers. Examples are papers that describe a previously unreported synthesis or reaction sequence, those that discuss structure–activity relationships, and those that describe compounds that are not very well-known. However, structures are not necessary just because a compound is being discussed, especially if you give the systematic name of the compound. Often, a line formula is adequate, such as CH_3COOH, C_6H_6, ClC_6H_5, or $1,4-Cl_2C_6H_4$. Do not supply art for material that can and should be portrayed on one line. Do not make angular a formula that could be portrayed linearly. Likewise, reactions need not be displayed if they can be explained clearly in text.

How To Prepare Structures and Schemes

As with illustrations, it is best to have a professional artist or draftsman prepare your structures and schemes. Keep in mind that the overall size requirements are the same as for illustrations, and that structures and schemes should also be drawn 50% larger than the final published size. Do not draw structures freehand. Use a template or transfer type (e.g., Letraset 4263 and 4264) and the same pens and paper as for line artwork.

The size of the rings and the size of the type should be proportional. The published size of six-membered rings should be approximately ⅜ inch (10 mm), five-membered rings should be about ¼ inch (6 mm), and the type size should be from 5 to 8 points. Therefore, if structures are drawn 50% larger, the six-membered rings should be approximately ½ inch (12 mm), the five-membered rings should be ⅜ inch (10 mm), and the type size should be from 9 to 12 points.

Oddly shaped rings should be consistent throughout a manuscript. The shapes of bicyclic structures also should be consistent. In multiring structures such as steroids, use partial structures that show only the pertinent points.

In three-dimensional drawings, lines in the background that are crossed by lines in the foreground should be broken to give a greater three-dimensional effect. Lines in the foreground should be made heavier.

Arrange structures in horizontal rows so that the display or block will fill a single-column width. Very large groups that require a double-column width in journals should be called Schemes (reaction diagrams) or Charts (groups of structures).

Center the compound numbers just below the structures. If you are discussing compounds of similar structure, draw only one parent structure with a general designation (e.g., S, R, or Ar) at the position where the substituents differ, and specify modifications below the structure with different numbers (see Figures 6A and 6B).

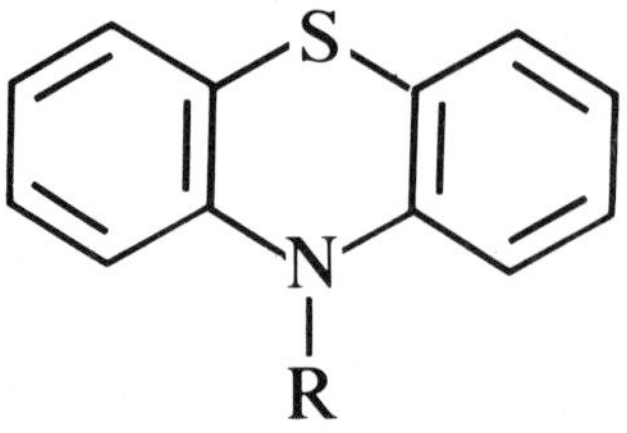

$$R = CH_3 \text{ or } C_2H_5$$

Figure 6A. Structure at the proper original size.

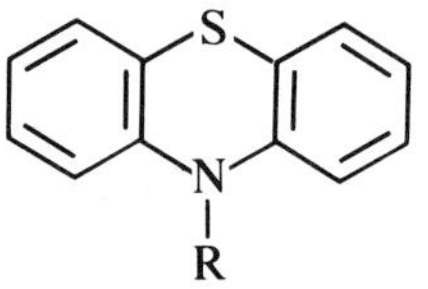

$R = CH_3 \text{ or } C_2H_5$

Figure 6B. Figure 6A as reduced for publication.

Do not place structures within tables unless you provide the entire table as camera-ready copy. Most typesetters cannot insert artwork into tables. It is best to provide structures as artwork, number them, and use only the numbers in the table.

Figures 7A–7D illustrate the following points.

The reaction arrows should be centered top to bottom in the structure height. The center for all structures is the center of the "tallest" structure on the same line. The centers of all structures should align.

The reaction arrows should be clear of the full width of the structures, including side chains.

Use the full column width before starting a new line. A compact presentation is most effective. Do not waste space, either vertical or horizontal. Schemes read from left to right, just like English sentences. As long as the proper sequence is maintained, it does not matter which line any given structure is on. If a reaction continues to the next line, it is best to have the arrow on the top line.

Do not place circles around plus or minus signs.

How To Submit Structures and Schemes

You should submit the original camera-ready structures with the original manuscript, and photocopies with each copy of the manuscript.

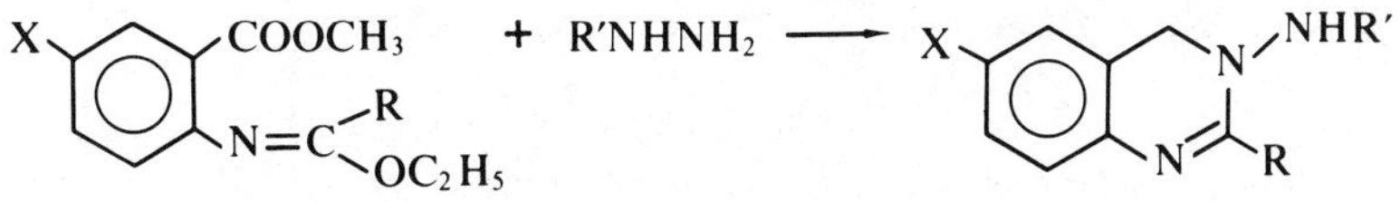

Figure 7A. Poor: *The arrow and the one-line structure are not centered from top to bottom in the height of the long structure.*

Figure 7B. Poor: *The reaction arrow is not clear of the full width of the structure.*

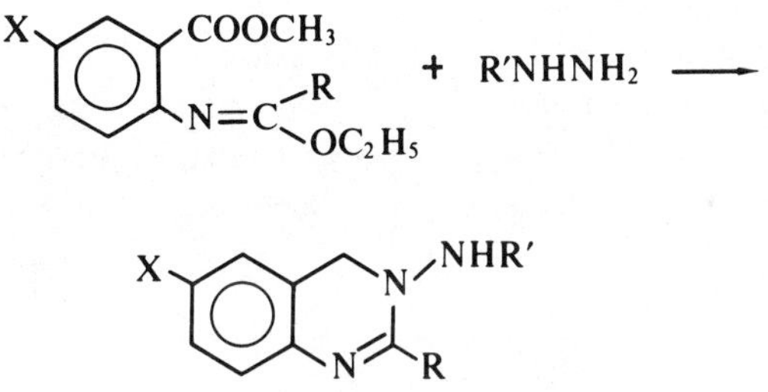

Figure 7C. Poor: *The full column width is not used; space is wasted.*

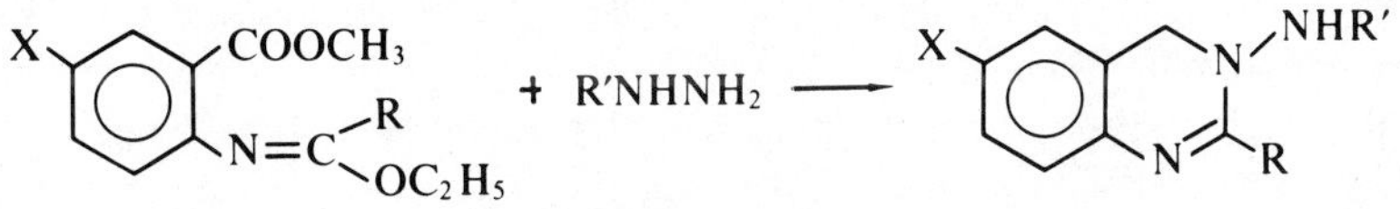

Figure 7D. Better: *The arrow and the one-line structure are centered from top to bottom in the height of the long structure; the reaction arrow is clear of the full width of the structure, including side chains; the arrow and the plus sign have an equal amount of space on both sides; and the full column width is used.*

How To Number and Cite Structures and Schemes

Number all structures with boldface numerals. If you discuss a compound several times in text, refer to it by its number only.

Simple one- or two-line reactions can be numbered the same as mathematical equations, that is, with a lightface Arabic number in parentheses at the right margin. Use one set of sequential numbers for both chemical reactions and mathematical equations. Do not confuse structure numbers with reaction numbers. Reaction schemes that include several reactions and require more than two lines should be numbered separately with Roman numerals and cited as "Scheme I", "Scheme II", and so on.

Tables

When To Use Tables

Use tables when the data are precise numbers that must be presented, when there are too many to be presented clearly as narrative, or when more meaningful interrelationships can be conveyed. Tables should supplement, not duplicate, text and figures. If you do not treat the data theoretically or if the material is not a major topic of discussion, do not present it in tables. Examples of material that is best handled as narrative in text are IR absorptions and NMR chemical shifts. In many instances, one table with representative data, rather than several tables, is all that is needed to illustrate a point.

How To Construct Tables

There are two kinds of tables: informal (or in-text) and formal. An informal table is one that consists of three to five lines and is no more than four columns wide. Informal tables may be placed in text following an introductory sentence. They are not given titles or numbers.

A formal table should consist of at least three columns, and the center and right columns must refer back to the left column. If you have only two columns, try writing the material as narrative. If you have three columns, but they do not relate to each other, perhaps the material is really a list of items and not a table at all. If your table has alignment and positioning requirements, perhaps it should really be a figure. It is important to understand these differences because tables are much more expensive to typeset than text; the larger the table, the more expensive. A well-constructed, meaningful table is worth the extra money, but anything else is a waste of money and does not enhance your paper.

Tables should be simple and concise, but if you have many small tables, consider combining some. Combining is usually possible when the same column is repeated in separate tables. Use consistent wording for all elements of similar or related tables. Use symbols and abbreviations that are consistent among tables and between tables and text.

Title   Every table must have a brief title that describes its contents. The title should be complete enough to be understood without referring to the text, and it should not contain new information that is not in the text. Put details in footnotes, not in the title.

Column Headings   Every column must have a heading that describes the material below it. Be as succinct as possible, keep headings to two lines, and use abbreviations and symbols. Define nonstandard abbreviations in footnotes. Name the parameter being measured and indicate the unit of measure after a comma. A unit of measure alone is not an acceptable column heading. If a general column heading applies to more than one column, use a rule below it and over the number of columns to which it applies; below the rule, give the specific headings for each column. A general column heading should not apply to the entire table; information that describes all of the columns belongs in a general footnote.

Columns   Usually, the leftmost column is the *stub* or *reading column*. All other columns should refer back to it. Stub entries should be consistent with the text as well as logical and grammatically consistent with each other. Main stub entries may also have subentries, which should be indented.

Material in columns can be aligned in four different ways: on the left, on the right, on the decimals, or centered. Words are usually aligned on the left, and numbers are usually aligned on the decimals. Use a zero before the decimal point. Groups of numbers connected by plus–minus signs and ranges of numbers connected by en dashes can also be usually aligned on the symbols and centered in the column width.

Do not use ditto marks or the word "ditto". Define nonstandard abbreviations in footnotes. Try not to have any one entry much longer than all the others. Place any explanatory material for specific entries in footnotes.

Be sure that all of the columns are really necessary. If there are no data in most of the entries of a column, it probably should be deleted and replaced with a general footnote. Or, if the entries in the entire column are the same, the column should be replaced with a footnote that says "in all cases, the value was x" or whatever is appropriate.

Footnotes   In footnotes, include explanatory material referring to the whole table and to specific entries. Information that should be placed in general footnotes referring to the whole table is the following: units of measure that apply to all entries in the table, explanations of abbreviations and symbols used frequently throughout the table, details of experimental conditions if not already described in text or if different from the text, sources of data, and other literature citations.

Examples of information that should be placed in specific footnotes are as follows: units of measure that are too long to fit in the column headings, explanations of abbreviations and symbols used with one or two entries, statistical significance of entries, experimental details that apply to specific entries, and different sources of data.

In some publications, general footnotes and sources are not cited with superscripts; they are labeled "Note" and "Source", respectively; specific footnotes are cited with superscripts. In some publications, all footnotes are cited with superscripts. Check the publication to which you are submitting your paper.

Use superscript, lower-case italic (underlined) letters, starting from the top and proceeding from left to right. Write all footnotes as narrative, in paragraph form, and use standard punctuation.

Table Width   The table width will of course depend on the widths of the individual columns. Very generally, tables having up to six columns will fit in a single journal column; tables having up to thirteen columns will fit in the double-column spread. Tables that exceed the double-column spread create difficulties in typesetting, so it is best to design tables to have less than eleven columns.

In books, tables having up to eight columns can fit the page width; tables having up to twelve columns will be turned broadside; larger tables can span two pages.

How To Submit Tables

Submit formal tables on separate pages after the reference section of the text, typed double spaced. The entire table must be typed double spaced. If the table must contain structures or other art or special symbols, or if the table has special alignment and positioning requirements, submit it as camera-ready copy, in which case it should be single spaced. If it is too large for the column or page, it will be reduced.

Submit informal tables in place in text, typed double spaced.

How To Number and Cite Tables

Number the formal tables sequentially with Roman numerals, in order of discussion in the text. Like figures, every table must be cited in the text.

Lists

Sometimes you may need to give numerous examples of items, such as chemical names. In such cases, if there are too many to run into text, they can be set as a list in some publications. A list of names is not truly a figure and not really a table. Give the list an unnumbered title.

Example

Potentially Carcinogenic Medicines

cyclophosphamide	melphalan
melphalan	chlorambucil
myleran	cisplatin
dacarbazine	procarbazine
chlornaphazine	phenoxybenzamine
treosulphan	thiotepa
methotrexate	5-fluorouracil
6-mercaptopurine	azathioprine
6-azacytidine	cytosine arabinoside
nitrofurazone	chloramphenicol
phenacetin	paracetamol
hexamethylmelamine	methapyrilene
diazepam	cimetidine

Reproducing Previously Published Materials

If you construct a figure or a table from data that were previously published in text, you do not need permission, but you should reference the source of the data.

If you significantly change a figure or table that was previously published, you do not need permission. Significant changes would be the addition of as much information as was in the original. You must change the original substantially to use it without permission. However, you should reference the original source and give proper credit.

If you wish to reproduce a previously published figure or table, consider carefully whether it is necessary to reproduce it or if citing it as a reference is adequate. Printing is very expensive and should not be unnecessary.

If you do need to reproduce a figure or table that has been published elsewhere, you must obtain permission in writing from the copyright owner (usually the publisher), and you must send the written permissions along with your final manuscript. Even if you were the author of the previously published figure or table, you still need permission from the copyright owner. The only exception is for a work of the U.S. government. Chapter 4 also discusses permission to reprint copyrighted materials.

4
Chapter

Copyright and Permissions

BARBARA FRIEDMAN POLANSKY

Definitions and General Copyright Questions

What Is Copyright?

Copyright is the exclusive legal right to reproduce, publish, and sell the matter and form of a literary, musical, or artistic work (*Webster's Third New International Dictionary,* Merriam Company, Publishers). According to Section 102 of the 1976 U.S. Copyright Law (Title 17, U.S. Code), which became effective January 1, 1978, copyright is automatically secured when original works of authorship are "fixed in any tangible medium of expression, now known or later developed, from which they can be perceived, reproduced, or otherwise communicated, either directly or with the aid of a machine or device." So, if you write a paper, compose a symphony, paint a picture, or take a photograph, you automatically own copyright, unless you did it for your employer or unless you were commissioned to do it. You need not fill out any forms or do anything further to secure your ownership of copyright.

The author is the original owner of copyright unless he or she transfers it in writing or unless the work is a "work made for hire". In that case, the employer or other person for whom the work was prepared is considered the author and owns all the rights, unless the parties have explicitly agreed otherwise in writing. Authors of a joint work are co-owners of copyright.

What Is a "Work Made for Hire"?

According to Section 101 of the 1976 U.S. Copyright Law, a "work made for hire is (1) a work prepared by an employee within the scope of his or her employment; or (2) a work specially ordered or commissioned...if the parties expressly agree" in writing that the work shall be considered a work made for hire. Under definition 1, if you have prepared a paper within the scope of your employment duties while working for the ABC Company, then the ABC Company owns copyright to your research paper. The company does not need to have a written agreement with you to claim copyright to papers you prepare within the scope of your employment. Copyright ownership should not be confused with any agreement concerning patent rights.

What Are the Exclusive Rights of a Copyright Owner?

According to Section 106 of the 1976 U.S. Copyright Law, the owner of copyright has the exclusive rights to do and to authorize any of the following:

1. to reproduce the copyrighted work in copies or phonorecords;
2. to prepare derivative works based upon the copyrighted work;
3. to distribute copies or phonorecords of the copyrighted work to the public by sale or other transfer of ownership, or by rental, lease, or lending;
4. in the case of literary, musical, dramatic, and choreographic works, pantomimes, and motion pictures and other audiovisual works, to perform the work publicly; and
5. in the case of literary, musical, dramatic, and choreographic works, pantomimes, and pictorial, graphic, or sculptural works, including the individual images of a motion picture or other audiovisual work, to display the copyrighted work publicly.

Can the Copyright Owner Give Any of the Rights to Another Party?

Copyright is divisible, so it is possible for the copyright owner to give someone the exclusive right to display the work, while keeping the other exclusive rights. A copyright owner can also grant nonexclusive rights. For instance, the copyright owner can grant a nonexclusive right to one company to make 1000 copies and then grant the same nonexclusive right to another company. A nonexclusive right is simply a right that does not exclusively belong to any one person or organization.

How Can Copyright Be Transferred?

Copyright transfers must be made in writing by the copyright owner. If the owner is an employer, an authorized agent of the employer must sign, unless the employer has given this right to its employees.

How Long Does Copyright Last?

In general, for authored works, copyright lasts for the life of the author plus 50 years. For commissioned works and works made for hire, the term of copyright is 100 years from creation or 75 years from publication, whichever is shorter. The ACS owns copyright to its published papers, so the ACS is considered the author and therefore the term of copyright is generally 75 years from publication.

When Is a Work in the Public Domain?

When a work falls into the public domain, people may use (reprint, republish, translate, etc.) that work without obtaining permission from anyone. The work is free to use in any manner.

A work is in the public domain if ALL authors are employees of the U.S. Government and have prepared the work as part of their official duties. Also, if a work was published before 1978 without a copyright notice, that work is in the public domain. Works also fall into the public domain when their copyrights expire.

What Is "Fair Use"?

The 1976 U.S. Copyright Law states that "fair use" of copyrighted material is not an infringement of copyright. Although "fair use" is defined in sections 107 and 108, the definition is very general and difficult to interpret. Because of the many interpretations that are possible, I will not define the term here. You should consult a legal counsel if you would like to know whether or not a particular use is fair. Use good judgment; when in doubt, it is a good idea to obtain permission from a copyright owner.

The following is section 107, "Limitations on exclusive rights: Fair use":

> Notwithstanding the provisions of section 106, the fair use of a copyrighted work, including such use by reproduction in copies or phonorecords or by any other means specified by that section, for purposes such as criticism, comment, news reporting, teaching (including multiple copies for classroom use), scholarship, or research, is not an infringement of copyright. In determining whether the use made of a work in any particular case is a fair use the factors to be considered shall include—
>
> (1) the purpose and character of the use, including whether such use is of a commercial nature or is for nonprofit educational purposes;
>
> (2) the nature of the copyrighted work;
>
> (3) the amount and substantiality of the portion used in relation to the copyrighted work as a whole; and
>
> (4) the effect of the use upon the potential market for or value of the copyrighted work.

Sources of Information

The best source of general copyright information is the U.S. Copyright Office (Register of Copyrights, Library of Congress, Washington, DC 20559). You can phone the Copyright Office's public information number, (202) 287-8700, during business hours.

The U.S. Copyright Office provides application forms, regulations, and information circulars at no cost. One of the most informative circulars is Circular R1, "Copyright Basics." You might also find these circulars helpful and useful:

R2	Publications on Copyright
R2b	Selected Bibliographies on Copyright
R4	Copyright Fees
R15a	Duration of Copyright
R21	Reproduction of Copyrighted Works by Educators and Librarians
R22	How To Investigate the Copyright Status of a Work
R23	The Copyright Card Catalog and the Online Files of the Copyright Office
R40a	Specifications for Visual Arts Identifying Material
R45	Copyright Registration for Motion Pictures Including Video Recordings
R56	Copyright for Sound Recordings
R61	Copyright Registration for Computer Programs
R96	(Section 202.1) Material Not Subject to Copyright
R96	(Section 202.2) Copyright Notice
R99	Highlights of the Current Copyright Law

The ACS also publishes papers on various aspects of copyright. Most of these papers, which are presented at ACS copyright symposiums and in general sessions held during ACS national meetings, are published in the *Journal of Chemical Information and Computer Sciences*. Many other sources of copyright information have been published and are available from local libraries.

How Copyright Affects Us: Specific Cases

The Copyright Act touches us almost every day. For instance, the song "Happy Birthday" is protected by copyright until the year 2010. If this popular song is republished, sung on TV or in the movies, or performed for the public, the "infringing party", if it has not already paid royalty fees, is asked to pay $25.00 per performance. If you own a digital watch that plays songs, including "Happy Birthday", you have already paid extra to cover the royalty fees.

Stores and restaurants that play music over loudspeakers are required either to obtain copyright royalty licenses to broadcast the music or to subscribe to a commercial broadcast music service. This decision was handed down by the Supreme Court on April 26, 1982, when it ruled that 525 clothing stores were infringing music copyright by playing the radio over loudspeakers as background music for their customers, unless they paid copyright royalty fees.

The Betamax case (*Sony Corp. v. Universal Studios*) is another famous copyright case that went to the Supreme Court. In 1976, Universal Studios sued Sony Corporation on copyright infringement grounds and charged that Sony's Betamax video cassette recorders were being used to tape copyrighted television programs. The Supreme Court found that home videotaping of copyrighted television programs for personal use is not an infringement and that Sony was not guilty of contributory infringement.

In late 1984, the Software Protection Act of 1984 and the Record Rental Amendment of 1984 were signed into law. Both laws require authorization from the copyright owner before one may rent, lease, or lend computer programs or records for commercial purposes.

More related to publishing is the case of New York University. In April 1983, NYU reported that it reached an agreement with nine publishers who had filed a lawsuit for copyright infringement in December 1982. The publishers charged that NYU, ten of its faculty members, and the Unique Copy Center violated copyright laws by photocopying textbooks and other educational material. The federal lawsuit was considered a test case to demonstrate the publishers' concerns that abuse by illegal photocopying was widespread on college campuses throughout the country and that copyright owners were suffering the financial consequences.

In an out-of-court settlement, NYU agreed to adopt specified guidelines on photocopying educational material, to request copying permission for material that is not covered by the agreement, and to pay royalty fees to the copyright owners, if requested. The guidelines do not pertain to students, who may copy articles for their own studying purposes. The guidelines permit teachers to make single copies of book chapters, articles, short stories, and other relatively short items for use in preparing their classes. Teachers may make multiple copies of copyrighted material, provided that the material is brief, that it will be needed before permission can be obtained, and that it will only be used for one semester.

American Cyanamid Company and E. R. Squibb Corporation were also sued by publishers of scientific and technical journals and other reference material; out-of court agreements were also reached in these cases. The publishers had charged that the companies were infringing the publishers' copyrights by making unauthorized photocopies. Under the terms of the settlements, which were very similar, the companies agreed to pay copying fees on (1) all copies of CCC-entered material (see next paragraph) made on central copying facilities and other attended copying equipment located on its premises; (2) copies received from outside sources, such as interlibrary loans or document-supply sources, unless notified that payment was made by the outside source; and (3) copies made on unattended copying equipment. The agreement with Squibb is different from the one with Cyanamid in that

Squibb may exclude from reporting up to 6% of their total copying and claim this amount as fair use.

To assist large industrial and other users in reporting their copying activity, in 1983 the Copyright Clearance Center (CCC) instituted the Annual Authorizations Service, which is a photocopy authorizations program that allows users to make copies of participating publications *without recording* or reporting internal copying activity, except during an audit period. Participating copyright owners sign an agreement with the CCC and give CCC the right to enter into a licensing agreement with users. Basically, the users make an annual payment to the CCC for the license to make all the copies they wish of articles included in the licensing program, provided that the copies are used for internal purposes at a corporate site during a one-year period.

For users other than large industrial organizations, the new Authorizations Service does not replace the CCC's transactional reporting system, in which users record all copies made.

ACS Copyright Policy

ACS copyright policy is governed by its bylaw on copyright, Bylaw IV, Section 1(b), which states that "for any writing of an author published by the Society in any of its books, journals, or other publications, the Society shall own the copyright for the original and any renewal thereof." Therefore, authors or their employers in any case of works made for hire (except authors of public domain works or works of a foreign government) are required to assign copyright to ACS for scholarly information that is submitted for publication.

Under the terms of the 1976 U.S. Copyright Law, the Society is required specifically to obtain copyright transfer from authors of each paper to allow the ACS to continue its past practice of holding copyright on individual articles and book chapters. This requirement is necessary for the Society to carry on its normal publishing activities of disseminating information, granting reprint permissions, making and supplying reprints of articles, collecting royalty fees or other fees, repackaging and distributing its published information, inputting printed material into its databases, and authorizing others to have any of these rights. In return for copyright transfer, the ACS grants back certain rights to authors or to their employers in the case of works made for hire.

ACS Copyright Status Form

The ACS Copyright Status Form is printed every year in the first issue of each ACS research journal; all authors of research papers in ACS journals and books are required to sign an ACS form.

COPYRIGHT STATUS FORM

Author(s)

Ms No.

Ms Title

Received

This manuscript will be published with the understanding it has been submitted on an exclusive basis.

Print or
Type
Author's
Name and
Address

RETURN FORM TO:

Editor
1155 — 16th Street, N.W.
Washington, D.C. 20036

COPYRIGHT TRANSFER

The undersigned, with the consent of all authors, hereby transfers, to the extent that there is copyright to be transferred, the exclusive copyright interest in the above cited manuscript (subsequently referred to as the "work") to the **American Chemical Society** subject to the following (Note: if the manuscript is not accepted by ACS or if it is withdrawn prior to acceptance by ACS, this transfer will be null and void and the form will be returned.):

A. The undersigned author and all coauthors retain the right to revise, adapt, prepare derivative works, present orally, or distribute the work provided that all such use is for the personal noncommercial benefit of the author(s) and is consistent with any prior contractual agreement between the undersigned and/or coauthors and their employer(s).

B. In all instances where the work is prepared as a "work made for hire" for an employer, the employer(s) of the author(s) retain(s) the right to revise, adapt, prepare derivative works, publish, reprint, reproduce, and distribute the work provided that all such use is for the promotion of its business enterprise and does not imply the endorsement of the American Chemical Society.

C. Whenever the American Chemical Society is approached by third parties for individual permission to use, reprint, or republish specified articles (except for classroom use, library reserve, or to reprint in a collective work) the undersigned author's or employer's permission will also be required.

D. No proprietary right other than copyright is claimed by the American Chemical Society.

E. For works prepared under U.S. Government contract or by employees of a foreign government or its instrumentalities, the American Chemical Society recognizes that government's prior nonexclusive, royalty-free license to publish, translate, reproduce, use, or dispose of the published form of the work, or allow others to do so for noncommercial government purposes. State contract number: _____ .

SIGN HERE FOR COPYRIGHT TRANSFER [Individual Author or Employer's Authorized Agent (work made for hire)]

Print Author's Name

Print Agent's Name and Title

Original Signature of Author on Behalf of All Authors (in Ink)   Date   Original Signature of Agent (in Ink)

CERTIFICATION AS A WORK OF THE U.S. GOVERNMENT

This is to certify that **ALL** authors are or were bona fide officers or employees of the U.S. Government at the time the paper was prepared, and that the work is a "work of the U.S. Government" (prepared by an officer or employee of the U.S. Government as a part of official duties), and, therefore, it is not subject to U.S. copyright. (This section should NOT be signed if the work was prepared under a government contract or coauthored by a non-U.S. Government employee.)

INDIVIDUAL AUTHOR **OR** AGENCY REPRESENTATIVE

Print Author's Name

Print Agency Representative's Name and Title

Original Signature of Author (in Ink)   Date   Original Signature of Agency Representative (in Ink)

FOREIGN COPYRIGHT RESERVED (NOTE: If your government permits copyright to be transferred, refer to section E and sign this form in the top section.)

☐ If **ALL** authors are employees of a foreign government that reserves its own copyright as mandated by national law, **DO NOT SIGN THIS FORM.** Please check this box as your request for the FOREIGN GOVERNMENT COPYRIGHT FORM (Blue Form) which you will be required to sign. **If you check this box**, mail this form to: Copyright Administrator, Books and Journals Division, American Chemical Society, 1155 Sixteenth Street, N.W., Washington, D.C. 20036, U.S.A.

The ACS Copyright Status Form

Rights Returned to Authors and Their Employers

The rights that ACS returns to authors and their employers appear on the ACS Copyright Status Form. Section A of the form states that "the undersigned author and all coauthors retain the right to revise, adapt, prepare derivative works, present orally, or distribute the work provided that all such use is for the personal noncommercial benefit of the author(s) and is consistent with any prior contractual agreement between the undersigned and/or coauthors and their employer(s)."

Section B states that "in all instances where the work is prepared as a 'work made for hire' for an employer, the employer(s) of the author(s) retain(s) the right to revise, adapt, prepare derivative works, publish, reprint, reproduce, and distribute the work provided that all such use is for the promotion of its business enterprise and does not imply the endorsement of the American Chemical Society."

Sections C and D are also very important for authors to note. The following statement appears under Section C: "Whenever the American Chemical Society is approached by third parties for individual permission to use, reprint, or republish specified articles (except for classroom use, library reserve, or to reprint in a collective work) the undersigned author's or employer's permission will also be required." Section D ensures authors and employers that "no proprietary right other than copyright is claimed by the American Chemical Society."

Authors Who Are Government Employees

Section E of the ACS form concerns contracted works of the U.S. Government or works prepared by employees of a foreign government or its instrumentalities: "[T]he American Chemical Society recognizes that government's prior nonexclusive, royalty-free license to publish, translate, reproduce, use, or dispose of the published form of the work, or allow others to do so for noncommercial government purposes."

The boxed section of the ACS form is for certification that a work is a "work of the U.S. Government", which is defined as a work prepared by an officer or employee of the U.S. Government as part of official duties. Such works are not subject to U.S. copyright. This certification statement is to be signed only when ALL authors are or were bona fide officers or employees of the U.S. Government at the time the paper was prepared. If at least one author was not an employee of the U.S. Government when the paper was prepared, that author should sign the top section of the ACS form, thereby assigning whatever copyrighted contributions that exist in the paper to the ACS.

The ACS recognizes that some foreign governments reserve copyright under their national laws. If all authors are employees of such governments, they will be asked to sign the ACS's Foreign Government Copyright Form, which gives to ACS all the rights it needs to carry on its publishing activities. Authors may request this form by checking the appropriate space on the Copyright Status Form.

Submitting a Properly Signed Form

Substitute forms or changes to the ACS Copyright Status Form are not acceptable. Any additions or changes made to the form will delay the processing of a paper for publication. An author, or the employer's authorized agent in the case of a work made for hire, must sign only one section of the form. The ACS form states that the bottom section is to be signed when ALL authors are employees of the U.S. Government. However, ACS policy is to accept a copyright form if both sections are signed, but only if the signatures are those of different authors. Also, an original signature or a stamped signature must be on the form; photocopied signatures are not acceptable.

These provisions are required so that the ACS can efficiently handle the more than 10,000 copyright forms that it receives each year. Most of the ACS publications assistants, who initially receive and process the copyright forms for the ACS's publications, are not located at ACS headquarters; they are located throughout the United States. Their primary concern is processing papers for publication. Although our publications assistants know that we require copyright assignment from authors, they are not familiar enough with copyright to know what changes would be acceptable. Therefore, it is efficient and effective for us to accept only the ACS form signed as is, without changes.

If ACS publications assistants receive any unacceptable forms, they will ask the author to submit another ACS Copyright Status Form or they will ask the ACS Copyright Administrator to handle the request. Usually, production on a paper may continue; *however, if a paper is ready for publication and we do not have a properly signed form, the paper will be delayed until an acceptable form is received.* Sometimes, though, an unacceptable form does slip through and a paper is still published. But even after publication, a thorough check is made of all copyright forms, and authors are asked to submit correctly signed forms when necessary.

Liability and Rights

Authors are solely responsible for the accuracy of their contributions. The ACS and the editors assume no responsibility for the statements and opinions advanced by the contributors to these publications.

Contributions that have appeared or been accepted for publication with essentially the same content in another journal or in some freely available printed work (e.g., government publications and proceedings) will not be considered for publication in an ACS journal or book. However, this restriction does not apply to results previously published as communications or letters to the editor in the same or other journals.

Manuscripts published in ACS books or journals and for which copyright has been transferred to the ACS may not be reprinted elsewhere in whole or in part without written permission from both the ACS and the authors. Authors who wish to reproduce their own articles for commercial use elsewhere also must have the consent of the ACS. Other reproduction is permitted only after obtaining the written consent of the ACS. (Requests should be addressed to the ACS Copyright Administrator [see page 150].)

Reprint Permissions

Material (tables, figures, charts, schemes, excerpts, etc.) that has appeared in a medium for which copyright is held by a person or organization other than ACS cannot be reprinted without the permission of the copyright holder. *It is the obligation of the author to secure this permission.* The author needs to obtain such permission even if the material appeared in an article that he or she originally wrote, and he or she transferred copyright. In addition, if the copyrighted material is from papers of other authors (rather than the author(s) who is submitting the manuscript in question), prior approval of that author should also be obtained.

When you request permission from the ACS to use ACS-copyrighted material in another medium, you should specify the journal name, volume number, year of publication, and the exact information you wish to use, such as entire article (give the inclusive page numbers of the requested material), Figures 3-5, Table II, etc., as well as how and where you wish to reuse the material. Also, include your requests for all ACS· information (even if you would like to use material from different ACS publications) on one sheet of paper, if possible. This format will speed the processing of your request.

Seeking Permission To Reproduce Papers Presented at ACS Meetings

When authors present papers at an ACS meeting, they own copyright to their own work, unless they have already transferred it or plan to transfer it to another publisher or to ACS for publication in an ACS journal or book. Authors may not grant permission to reproduce their own papers if they have already transferred copyright to a publisher. Furthermore, *authors themselves may not republish their own papers in whole or in part if they have already transferred copyright to a publisher, unless* the publisher allows this right. However, if an author still owns copyright to his or her work, you should contact the author for permission to reproduce the work. If the author grants permission, you need not obtain further permission from the ACS. You should acknowledge, however, that the work was presented at an ACS meeting and that it was published as an ACS preprinted work.

If an author has already submitted or intends to submit his or her paper to an ACS journal or book, permission requests should be directed to the ACS Copyright Administrator. Such requests should also mention the publication in which the work will appear.

If an author's paper appears in an ACS division's preprint publication, you should contact the author first to determine whether he or she has transferred copyright in writing to the ACS division. Some divisions (e.g., Rubber Chemistry and Technology) do require copyright transfer, so it is mandatory to contact either the author or the division to obtain the necessary permission to reuse an author's work. An author can present a paper at an ACS meeting, the preprint can be made available by an ACS division, and the full paper can be published by another publisher.

Seeking Permission To Use Photographs

If you have a photograph of a person, you must have written permission from the person to reproduce the photograph. Likewise, if you have a photograph of commercial equipment, you must have written permission from the manufacturer to reproduce the photograph.

Copyright Clearance Center

The ACS does not grant blanket copying permission, unless you have a license-to-copy agreement with the ACS or unless you report your personal use or internal use copying activities, or those such activities of your specific clients, to the Copyright Clearance Center, Inc. (27 Congress Street, Salem, MA 01970).

The ACS is a registered publisher with the CCC. On the first page of each ACS-copyrighted journal article and book chapter is a long numeric code, referred to as the CCC code. The appearance of this code indicates ACS's consent that copies of articles or book chapters may be made for personal use or internal use, or for personal internal use of specific clients. The consent does not extend to other kinds of copying such as copying for general distribution, for advertising or promotional purposes, for creating new collective works, or for resale.

Copyright Credit

The ACS grants permission to republish or reprint ACS-copyrighted material provided that requests are received in writing; that requesters agree to pay royalty fees, if any are due; and that the required ACS copyright credit line is used. For a journal, the standard credit line is "Reprinted with permission from (full journal reference). Copyright (year) American Chemical Society". For a book, a sample credit line is "Reprinted with permission from Whitehurst, D. D. In *Organic Chemistry of Coal*, Larsen, J. W., Ed., ACS Symposium Series No. 71; American Chemical Society: Washington, DC, 1978; p 13. Copyright 1978 American Chemical Society."

Guidelines for Classroom Use

According to the ACS guidelines for classroom use, ACS will grant royalty-free permission to make copies of journal article(s) or book chapter(s) provided that

1. ACS receives a written request listing each article to be copied and the number of copies to be made;

2. the required copyright credit line appears on the first page of each ACS-copyrighted article: "Reprinted with permission from (full journal reference). Copyright (year) American Chemical Society";

3. the article is used as a current teaching tool and will not be reused by the teacher in subsequent semesters;

4. the reprint or photocopy is not attached to any other article copies or material;

5. no charge is made to the student beyond the actual cost of the photocopying; and

6. use of the article(s) will not imply ACS endorsement of the course taught or any material contained in the requester article(s).

If your copying meets all but criterion 3, permission will be granted provided you agree in writing to pay the requested royalty fee for classroom use; please write to the the the ACS Copyright Administrator. Alternatively, you may report your copying to the Copyright Clearance Center.

If your copying does not meet criterion 4, 5, or 6, you must write to the ACS Copyright Administrator to make arrangements for a copying agreement.

Questions about ACS Copyright

Write or call the ACS Copyright Administrator, Books and Journals Division, American Chemical Society, 1155 Sixteenth Street, N.W., Washington, DC 20036. The direct telephone number is (202) 872-4367.

5
Chapter

Manuscript Submissions in Machine-Readable Form

MARIANNE C. BROGAN

With the advent of word processors and minicomputers, the role of authors in the publication process has taken on a new dimension— one of greater responsibility and autonomy. The age of the electronic manuscript is at hand, and with it come great opportunities for improvement in quality and timeliness of publication and for different approaches in information transfer.

In 1984, 75% of authors submitting papers to ACS publications had the equipment to produce machine-readable manuscripts; in 1984, 0.1% of all papers were submitted and processed via electronic media. The huge discrepancy between the possibility and the reality is attributed to the practicality of the procedure: It is no routine matter to overcome the difficulties associated with coding and transmitting scientific text, receiving that text and reencoding it, and producing quality publications in an economic fashion.

Technical difficulties abound, all surmountable, but at a cost—to the author or to the publisher. The questions of media, format, and coding give rise to many different answers. When the number of authors, publishers, and composition houses is added to the equation, the complexities increase tremendously.

The American Chemical Society can address one audience—the authors who publish in ACS books, journals, and magazines—and provide that audience with guidelines and protocols for electronic submission, including coding and flagging schemes. The ACS has such coding and flagging schemes, but their utilization is not a straightforward matter and their implementation would be applicable only within the ACS. Because ACS authors publish elsewhere, an ideal would be general guidelines, guidelines authors could use for both ACS and non-ACS publication.

The Association of American Publishers (AAP) is sponsoring a study that has as one of its objectives the achievement of that ideal: general guidelines. ACS is cooperating in their work. The following recommendations based upon the AAP approach would be generally applicable for all publishers willing to accept papers in electronic media.

Background

Before submitting a paper in electronic form, you must ask several questions: Is the publisher willing to accept a paper in electronic form? If so, in what medium? What coding instructions are necessary?

Thus, step 1 involves contacting the publishers, who will generally provide specific guidelines depending upon their capabilities and those of their composition houses.

Electronic information can be communicated in three standard ways: telecommunications, disk (diskette), and magnetic tape. Each has advantages and drawbacks:

Telecommunications: Established protocols but potentially costly for long papers and subject to transmission errors; may require special access codes or times to offset costs; will require telecommunications capability on the word processor or personal computer used in manuscript preparation.

Disk (Diskette): Device-dependent and therefore subject to compatibility problems (format, density, size considerations; data structure); will generally have to be converted to a disk or other electronic version that is supported by the publisher or composition house; conversion drives up costs. The publisher may restrict the options to disks from a few systems, with standard file storage.

Magnetic tape: Generally unwieldy; again subject to restrictions (e.g., record length, speed) but much more standardized, and therefore more transportable, than disk.

Having ascertained that the publisher will accept the paper in electronic form and the appropriate communications medium, you need to obtain specific instructions on procedures and coding.

Manuscript Preparation

Organization is a vital component in manuscript preparation. Good organization lends itself to efficient translation in electronic form. The recommendations in Chapter 1 are an effective starting point for the machine-readable file as well as the printed version. After the manuscript type and structure are determined and the paper is written, the manuscript components may be "tagged". If the paper is written from an outline, the tagging could be introduced concurrently with text generation. However, it is best to maintain two separate versions of the electronic file: one for initial submission and revision, lacking the coding schemes, and one for final submission, with the necessary coding incorporated.

Manuscript Input

Because a paper submitted for publication involves so much creative work, the temptation exists to make the manuscript resemble the format of the anticipated publication as much as possible. For electronic submissions, simplicity is the keynote: be consistent; be precise; avoid spaces, carriage returns, and tab commands; do not use any coding for boldface or italic for format purposes; and avoid using coding for boldface and italic in text. Unnecessary or incorrect codes must be removed; relevant codes must be inserted by the editing or production staff. Most format instructions will be table-generated and accessed by the system computer.

Use the following guidelines:

- Remove all word processor induced formats from the text file.

- Set all material flush left (do not use the tab key for any text component).

- Treat all identifiable manuscript components (titles, authors, address, abstract, etc.) as separate paragraphs.

- Use blank lines to separate paragraphs.

- Use only one space between sentences.

- Do not use l's (ells) for 1's (ones) or o's (ohs) for 0's (zeros).

- Be unambiguous and consistent with command codes.

Tagging (Use of Command Codes)

A "tag" is a statement of identification or equivalency. Tags are used for several purposes; data base applications, format instructions, and marking simplification are important considerations. For example, if an author data element is defined, that data element may be used for various applications: in the author field of the paper, in the table of contents, in the issue author index, in collective indexes, in running heads, and in whatever other applications may be subsequently devised. The author names could be transmitted in whatever type face, point size, fashion (inverted or not), and medium (printed, electronic) chosen for the application.

Among the important data elements, with the AAP-recommended codes, for book chapters and journal papers are the following:

Element	Keyboarding Convention
title	<ti> or <TI>
author	<a> or <A>
address (affiliation)	<af> or <AF>
head level 1	<h1> or <H1>
text level 1	<tx1> or <TX1>
chapter	<c> or <C>
acknowledgments	<ak> or <AK>
abstracts	<ab> or <AB>
footnote reference	<ff> or <FF>
footnote	<fn> or <FN>
bibliographic reference	<rbr> or <RBR>
figure	<fg> or <FG>
figure reference	<fr> or <FR>
table reference	<tr> or <TR>
matrix/table	<tb> or <TB>
paragraph	<p> or <P>
emphasis, type 1	
start	<el> or <El>
end	</>

Other data elements that are specific for ACS applications are discussed later.

Most commonly available word processor and computer terminal keyboards in the United States conform to ASCII standards. ASCII is a subset of ISO 646 IRV (International Standards Organization 646 International Reference Version) with two exceptions: general currency sign ($ in ASCII) and overline (tilde in ASCII). The AAP standard has adopted the ISO version as its basic character set, with the following characters and symbols:

upper case A through Z
lower case a through z

numerals 0 through 9
space
exclamation mark
quotation mark
greater and less than signs
percent sign
apostrophe
left and right parentheses
left and right square brackets
left and right curly brackets
asterisk
number sign
comma
ampersand

hyphen
full stop, period
solidus (slash, /)
reverse solidus (backslash, \)
colon
semicolon
equals sign
question mark
commercial at (@)
circumflex accent
underline
grave accent
vertical line (|)

All characters not part of this basis set require additional coding. Thus, Greek letters, mathematical symbols, superscripts, subscripts, and other special characters require explicit coding instructions. In addition, some of these characters (e.g., [, &) are used as control characters (in the definition of special characters or fonts); as such, their standard appearance must be triggered by appropriate codes. The AAP standard has addressed coding for letters of the non-English alphabet and will recommend mnemonic codes for other special characters, but many specific codes have not yet been established. The next section contains a list of ACS-defined codes for some of these characters.

Because Greek letters are such an important component of scientific text, the AAP coding scheme for Greek characters is given: for individual Greek characters the codes follow the form &G* (where * is the Roman alphabet character equivalent to the desired Greek character; e.g., &Ga is lower-case Greek alpha, &GF is upper-case Greek phi, &Gg is lower-case Greek gamma, &GD is upper-case Greek delta).

ACS-Specific Guidelines

Description

ACS, like most publishers, is capable of accepting papers in electronic form. Whether a specific paper is appropriate for electronic submission is, however, another matter. Factors such as complexity, cost, and timeliness enter into the decision.

On the surface, the larger the paper, the better the candidate for processing via electronic media. Thus, book chapters, reviews, and feature articles appear to be especially suitable for this technology. However, extensive mathematics, special character set demands, frequent chemical structures and other artwork, and special format considerations all create problems, both in editing and in composition.

Resolving the problems may drive up production costs, so that the apparent economy of capturing author keystrokes is imaginary and relief from the burden of proofreading is short-lived. Experience has shown that reading, modifying, translating, and editing electronic files, however communicated, may be more costly than re-creating and proofing those files by the staff of the composition facility.

ACS material is complex. The coding requests, especially because they are inserted after the manuscript is submitted, are such that an author may prefer to proofread a reset manuscript rather than ensure correctness of the necessary codes. A miscoded manuscript may create chaos, which can perhaps be restored to order only by conventional rekeying and thus would negate the additional work incurred by the author.

Manuscript Submission: The Review Process

Hard copy (the traditional paper copy) will continue to be the medium for initial submission and review. Only after revision and acceptance of the paper should a corrected electronic version of it be forwarded to editorial staff. Corrected paper copies will continue to be necessary. One of these paper copies will be forwarded by the editor to the editorial production office. Its receipt by that office will signal approval to process the electronic submission.

Manuscript Submission: The Production Process

ACS is prepared to accept manuscripts via telecommunications, on 9-track 1600-bpi (or higher) magnetic tape, or on 5¼-inch IBM PC and IBM AT diskettes (for journals and books) and DEC PC and Osborne diskettes (for books) if most of the necessary character codes are embedded. Specific details (establishing a "mail box" for phone access; record size and additional instructions for tape; restrictions or instructions for diskettes) may be obtained from the appropriate Washington or Columbus editorial office.

The AAP standard character set and standard coding scheme, supplemented by ACS recommendations for additional characters and data elements, should be used. ACS journals have a single composition facility, and hence a uniform format. Therefore, for ACS journals, the following special characters and data elements may be used:

Special Characters

Element Name	Code
subscript	$
sub subscript	$$
sub superscript	$#
superscript	#
super superscript	##
super subscript	#$
script alphabet	/sA (e.g., for script A)
boldface	— (underline)
italic	*
plus	/+
less than	<l
greater than	<g
dagger	<f
degree mark	>
section mark	<<sec
paragraph mark	<<par
en dash	-
left (reverse) arrow	<<rar
right (forward) arrow	<<far
double-headed arrow	<<dar
double bond	<<dbd
minus sign	-
multiplication sign	<<tme
product or centered dot	<<cdt

Data Elements

Element Name	Code
received date	<rcv>
revised received date	<rvd>
accepted date	<acd>
supplementary material	<sm>
meetings paragraph	<rcp>
title footnote	<caf>
subsidiary text	<tx2>
sentence	<sen>
start of equation	<<eqn
end of equation	<<edq##
equation placement	<<plr read in
	<<pla as soon as possible
	<<plj joined
equation type	<<mth math
	<<chm chem

These special codes generally precede the character(s) affected; for example, boldface 1 would be coded as __1

Example of a Coded Manuscript

<ti>Publication in Scientific Journals of Manuscripts Prepared via Electronic Methods
<au>Marianne Brogan
<af>Books and Journals Division, American Chemical Society, Columbus, OH 43210
<rcv>January 14, 1985
<ab>When manuscripts have been prepared by use of electronic media, direct use by the publisher of the initial keystrokes should result in significant savings, among these being proofreading time and reduction of duplication of effort and proofreading expense on the part of the publisher. However, these benefits will accrue only if the paper is prepared in a form suitable for easy manipulation on the part of the publisher. The problems associated with this condition are discussed, with special emphasis on scientific publications and data base applications.
<h1>Introduction
Composition of scientific material has long been a challenge. Wide use of special characters, italic and boldface type, superscripts and subscripts, and mathematical and chemical display equations augments the problems faced by publishers in other fields. Consider the following simple equation:

##equ# 1 <<eqn<<plr<<chmC$6H$5CHO /+ H$2NC$6H$5 = C$6H$5CH<<dbdNC$6H$5<<edq##

If the relatively simple coding used to indicate a display equation (##equ# 1) does not bother a prospective author, the coding used to generate the equation itself (<<eqn, start of equation; <<plr, read-in placement of equation; <<chm, type of equation; $6, subscript 6; etc.) may.

Flagging (for figures, tables, schemes, charts, footnotes, references) is necessary, as of course is specific identification of all manuscript components. For authors willing to face these challenges, specific instructions will be made available.

In addition, because much material published via modern technology will become part of an electronic data base, further search assistance tools are often useful. Thus, whereas the AAP standard does not focus on special sections such as Registry Number and Supplementary Material, ACS papers do. Hence, ACS must provide tags for the corresponding data elements.

References and footnotes again require special coding (for identification purposes) and flagging#1<ff1> (for placement). In the example just given, the placement code (<ff1>) should ensure the footnote's being placed at the bottom of the column containing this reference number. The superscript code (#1) will result in the 1 being placed in a superscript position. In some journals, the reference number is given in italic type, in parentheses [e.g., (*1*)]. In this case, the appropriate representation would be (*1) or (<E2>1</>), the former corresponding to the notation for italic in ACS publications and the latter corresponding to emphasis type 2 in the AAP standards (defined as italic for ACS). Because in these journals all references are collected at the end of the paper, no placement code is necessary.

<fn1>"The Association of American Publishers Author's Guide to Electronic Manuscript Preparation and Generic Tagging. Module 1: The Basics"; The Association of American Publishers, Inc.: Washington, DC, 1985.

6
Chapter

The Literature
Becoming Part of It and Using It

ARTHUR A. ANTONY

This chapter has basically two aims: (1) to help you have a better understanding of how your published works will be handled by the producers of information retrieval tools and (2) to help you find pertinent information by informing you about the major information retrieval resources.

We shall look first at the formats for publication that chemists and chemical engineers use to communicate the results of their work. Then we shall consider the information retrieval tools that announce those publications and provide intellectual access to their contents.

The Author's Role in Information Retrieval

After I accepted the assignment to prepare this chapter, I wrote to several major chemical secondary services and asked for their insight on how an author might write to optimize the information retrieval process. I received several responses that were very helpful.

The respondents in general concurred on the importance of phrasing the title. The title is nearly always the first piece of information that an indexer sees, and it will have considerable influence on how the document is handled. For some highly selective services, the title may be a significant factor in the decision to cover a document or not. Some printed indexes, such as *Chemical Titles*, rely on title words for keyword indexing; and for nearly all online data bases, title words are among the searchable parameters. Because not all secondary services are able to handle chemical formulas in titles, names are preferable to formulas.

The secondary services are particularly anxious for authors to prepare informative abstracts. *Chemical Abstracts* guidelines indicate a preference for abstracts that state the problem or purpose of the research, indicate the theoretical or experimental plan, accurately summarize the principal findings, and state the major conclusions.

Within the body of the text, the author should make clear what is new, and should unambiguously separate speculation from observation. Adequate referencing of antecedents in the literature also helps the indexer in determining what is new and what is not. It is important to be aware that a paper must be understood by the abstractors and indexers who handle it. Although the abstractors and indexers usually have very good technical backgrounds, they cannot be familiar with all of the jargon and specialized abbreviations used by a small group of specialists in close contact with one another. Thus, jargon and specialized abbreviations, if used, should be explained.

Secondary service respondents also advise authors to use the most specific terminology that applies. If an author uses "desulfurization", but is more specifically discussing "hydrodesulfurization", the document is likely to be indexed only by the more general concept, and might be overlooked by someone looking only for the more specific concept.

Secondary service indexers are on the lookout for stated safety and health hazards. Authors should highlight safety hazards in the experimental sections of their papers by describing them in separate paragraphs introduced by a word like "Caution".

The important role that abstracts, indexes, data bases, and libraries play in scientific progress cannot be overemphasized. It is therefore important that authors use these resources effectively and write so that these resources can be prepared for others to use effectively.

Primary Versus Secondary Literature

Primary publications are those in which information, data, and ideas are first presented to the public, whereas secondary resources either review and summarize the contents of primary publications, direct the information seeker to those publications, or both. The distinction is not always sharp. Primary publications nearly always contain some review information. In fact, they should, because chemistry is a cumulative science, in which new developments always have antecedents, and the author of a primary publication is expected to acknowledge those antecedents. Thus, a primary publication often proves to be a rich resource for leads to earlier publications in a search for information. Secondary publications, especially review papers and monographs, often express new ideas or insights as a consequence of bringing together previously unrelated facts. New data are almost never introduced in secondary publications.

Studies by Garvey and his co-workers (1) resulted in a model of scientific communication behavior in which a typical piece of scientific work progresses through eleven stages from its initiation to the preparation of a publication. The nature of the information that a scientist needs and the types of resources that can best provide it vary from one stage to the next (2). For example, scientific journals prove extremely useful during the formulation of a scientific solution to a problem (an early stage) and the "integration of the findings into the current state of scientific knowledge" (i.e., the latest stage, which generally involves preparing a formal publication). The journal apparently is much less useful during intermediate stages, when informal sources, especially trusted colleagues, are dominant.

Prior to formal publication, scientists often present some of their research to limited groups of colleagues, usually orally. Such oral presentations do not constitute "publication" and often result in modifications and improvements before formal publication is attempted. Informal written communications to distant colleagues engaged in similar research may also take place. The term "invisible college" has been applied to the networks that scientists establish among themselves for facilitating informal communication (3). The value of invisible colleges in fostering scientific progress should not be underestimated.

When a scientist publishes research results formally, in a journal or other primary medium, those results become part of the open literature, available to any scientist or engineer who is interested in them. This open publication means that the ideas expressed have a better chance of influencing future developments in science and its technological applications. However, retrieval of information directly from the primary literature is difficult, so secondary services have been established to facilitate that retrieval.

The first indexes to cover a primary publication are relatively uncomplicated indexes, usually based on author names and title words. Detailed subject analysis and the preparation of abstracts take more time, so the indexes and abstracts that provide these features usually appear later, but they have more lasting value for the future generations of scientists seeking information. The computer has helped speed up secondary processing and has itself become a useful tool for information retrieval via online data bases.

At any given time, a body of fundamental concepts called "paradigms" is universally accepted by the practitioners in a scientific field (4). The atomic–molecular model, for example, is one of the paradigms of chemistry. Three forms of publication expound the details of the paradigms of a field: treatises, encyclopedias, and textbooks.

The treatise is usually for the specialist, and chemistry is such a large field that there is no contemporary treatise for all of chemistry. The significant chemical treatises tend to be multivolume sets on fairly broad topics, authored (or more usually edited) by very eminent chemists. Examples of treatises include *Comprehensive Chemical Kinetics*, edited by C. H. Bamford and C. H. M. Tipper (5), and *Comprehensive Inorganic Chemistry*, edited by J. C. Bailar et al. (6).

Encyclopedias are arranged as reference books, usually with entries in alphabetical order, and they are usually intended for the nonspecialist. Textbooks, of course, are designed for educating newcomers to the field.

Journal Articles

To most chemists, "primary literature" means journal articles. The scientific journal has been a vehicle for transmitting information in chemistry and other sciences since 1665. The scientific journal format has been extremely successful, as evidenced by the exponential growth rate in journal publishing (7) and by the fact that at least some scientific journals are published in nearly every country. The results of chemical research have appeared in general scientific journals since their inception and still are sometimes published in journals such as *Nature* and *Science*, which encompass a broad spectrum of scientific disciplines.

Chemists usually publish in chemistry journals. Indeed, many specialized journals accept articles only in very narrow areas of chemical research. Still, prestigious general chemical journals, such as the *Journal of the American Chemical Society*, continue to attract very significant contributions.

Estimates of the number of scientific journals now being published range from around 25,000 to nearly 50,000 (8). In what is probably the most complete current listing of chemical journals, *Chemical Abstracts Service Source Index* (CASSI) identifies 18,155 "serials currently published". Many of these serials are primary chemistry or chemical engineering journals, but included in that number are review journals, interdisciplinary journals, journals from nonchemistry disciplines that sometimes publish articles of chemical interest, and serially produced publications that are not really journals at all. In any case, CASSI and the quarterly supplements that augment it are excellent sources for identifying journal titles of interest to the chemistry and chemical engineering communities. Perusal of a few randomly selected pages from CASSI indicates the variety of languages, countries of publication, and subdisciplines encompassed by primary journals in chemistry.

The chemist or chemical engineer may choose to publish primary information in some format other than the journal article. The dissertation, one form of primary literature, is usually a required publication of a prospective Ph.D. candidate, and dissertations are expected to be reports of original research. For chemical inventions, patents provide legal protection that cannot be achieved with other forms of primary publication. Patent-granting agencies insist on an invention's novelty.

Patents

A patent is a legal document in which an inventor discloses the technical details of his or her invention in return for the right to exclude others (i.e., potential competitors) from practicing the invention for a restricted time period. Some countries also have other restrictions.

The U.S. Patent and Trademark Office and similar agencies in other countries examine patent applications and grant and publish patents. An inventor may assign patent rights to another party, called the assignee. Usually the assignee is the company that employed the inventor and supported the research.

A U.S. patent is first announced to the world in the U.S. Patent and Trademark Office's *Official Gazette*. Each issue of the *Official Gazette* lists the patents issued that week, each with at least one claim, arranged in patent number order. Each issue is indexed by patentee and assignee and by patent classification number derived from an elaborate classification scheme. The classification number index provides the only direct subject access to the *Official Gazette*.

Patent offices in other countries also publish official listings of patents. An invention is often patented in more than one country, to provide international coverage. The variation in legal and other details from one country to another may be quite complex (9). In many countries, but not in the United States, a patent application meeting rather minimal requirements is opened to the public at a set time after the date of filing, without formal examination for novelty or other features. In the United States, the application is kept confidential by the patent office until the patent is issued. Patents usually take about two years to be issued. However, some U.S. patent applications that name a U.S. government agency as assignee are not kept confidential, but are announced in a government publication.

The complications associated with all the different patent documents that may be issued for a given chemical invention in different countries have been discussed in detail by Grubb (9). Here, we are concerned with how to identify patents that deal with a subject of interest, because patents contain a wealth of technical and scientific information. Thus patents are a part of the scientific literature. They may be read for background information, experimental details, or data; in effect they contribute to the development of science just as journal articles and other primary publications do. Hence, patents are cited not only in other patents but also in review articles, books, primary journal articles, etc.

The *Official Gazette* is a useful information retrieval tool for patents, but patents are so important in the commercial world that numerous specialized searching tools have been developed. The most important of these for the U.S. chemical community are discussed later.

Conference Papers

Robert Olby's book, *The Path to the Double Helix* (10), suggests how significant conferences often stand out as milestones in the minds (and conversations) of scientists when they reminisce about their careers. However, not everything that is presented at a meeting is recorded, nor does it deserve to be.

Our main interest here is in the formal presentations of research results at the national and international meetings in chemistry. Often at such meetings, a scientist first presents his or her research to the public at large; hence publications that emanate from such meetings are rightfully considered primary. However, some papers, especially those by scientists who have been invited to give presentations, are state-of-the-art reviews, that is, summaries of the most important work to date on a specific topic.

The research presented at large national meetings has usually been completed very shortly before the presentations (11). Consequently, many scientists consider attendance at such meetings to be vital to their ability to keep informed in their research areas.

The literature resulting from conferences is not as easy to locate as patent or journal literature. A great deal of the material is eventually published in journal articles, but not all of it. For example, researchers (12) have found that within two years almost half the material presented at several large national meetings in a variety of disciplines had appeared in journal articles.

Some meeting sponsors do not publish conference proceedings, presumably in anticipation that the authors will submit the more significant work to refereed journals. However, in many cases, conference proceedings do appear. Although the papers have probably not been as rigorously reviewed as those in the better journals, published proceedings receive a wide readership for at least three reasons: (1) the proceedings are sometimes in print long before the research results show up in published journal articles; (2) the conference proceedings gather all the papers in one place to form a unified "snapshot" of that particular conference's contributions to the field; and (3) many of the papers will never be published elsewhere.

The ACS publishes the abstracts of all the papers for each of its national meetings and many of its regional meetings in a continuing publication entitled *Abstracts of Papers*. Although proceedings of an entire meeting are not published as a unit, symposia on special subjects may be published by the ACS Books Department as part of the ACS Symposium Series or Advances in Chemistry series. Some ACS divisions (e.g., Environmental Chemistry, Fuel Chemistry, Polymer Chemistry) publish preprints of papers presented at their technical sessions. Most published conference papers are not verbatim transcripts of what was said. Verbatim transcription usually leads to very poor reading material, so authors prepare edited, but closely corresponding, versions of what they said.

Technical Reports

In terms of publication volume, technical reports constitute a major body of contemporary scientific primary publication. Scientists working under contracts or grants may be required to submit written progress reports to the agencies supporting their research. Others, not necessarily required to do so, may prepare written reports to disseminate information about work they are doing. Reports submitted to government agencies usually are considered part of the public domain and are given wide dissemination, unless they are classified as secret or are otherwise restricted.

The National Technical Information Service (NTIS) in Springfield, VA, is a major depository and distributor for technical reports. The principal vehicles for disseminating information about NTIS reports are the publication "Government Reports Announcements" and the NTIS data base.

Most NTIS reports have not been subjected to rigorous refereeing or editorial processes. For this reason, scientists occasionally label the reports as "worthless". That judgment is unfair because for the most part the reports have not been judged one way or the other. NTIS reports emphasize applied science and engineering and often contain technical details of vital importance to society. Patent applications that name a U.S. government agency as assignee are among the many different kinds of documents that are handled by NTIS as technical reports. This case is the one exception to the rule that U.S. patent applications are kept confidential until the patent is issued.

Dissertations and Theses

Dissertations and theses are documents that are prepared to partially meet the requirements for degrees beyond the Bachelor's degree. Original research is an almost universal requirement for the Ph.D. dissertation. Although the requirements are not as rigorous for a Master's degree and many Master's programs do not require a thesis, most Master's theses do report some original research results.

Most of the important research that chemistry dissertations describe eventually appears in other forms, usually as journal articles. Bottle (*13*) analyzed the publication patterns of dissertations in chemistry and showed that, on the average, a significant time lag occurs between the publication of a dissertation and the appearance of the information in journal form. The introductory subject review, extensive literature review, and concluding summary that are parts of most dissertations are rarely published elsewhere.

University Microfilms of Ann Arbor, MI, deserves credit for making Ph.D. dissertations widely available. Now, nearly all students receiving a Ph.D. in the United States and Canada, and (only in more recent years) some students in Western Europe, must have their dissertations filmed by University Microfilms, which, in turn, is able to offer copies for sale. The dissertations are listed and abstracted in *Dissertation Abstracts International*. A corresponding data base is also available for computer searching. Master's theses are usually much more difficult to identify and often are not available at all.

Review Literature

Some scientific journals specialize in publishing reviews. A good review collects a large amount of related information in one place, and therefore, helps the reader get an overall picture of a topic. At the same time, a review provides the reader with references to primary sources and possibly to other reviews. The list of references may be comprehensive or selective, depending on the intentions of the author and the amount of literature on the topic under review.

Woodward (*14*) suggests that reviews serve both "historical" and "contemporary" functions. The historical functions contribute to the development of science, whereas the contemporary functions benefit the individual scientists. Historical functions include the role of the review journal in providing a second level of peer evaluation. That is, the review journal takes the refereeing process one step further by recording the significance of a contribution. Other historical functions include the achievement of a unified picture of the state of knowledge on a topic, the compaction of knowledge by leaving out many of the details that are essential in the primary literature, and the suggestion of new areas for research. The individual scientist benefits from the contemporary functions in a number of ways: The scientist can concentrate his or her energies on reading and studying the most worthwhile contributions on a topic; timely reviews can help the scientist keep current on topics that are not at the core of his or her research interests; a review paper may help ensure that a scientist has not missed anything significant in the normal course of keeping up with the primary literature; the science student may be educated by a good review.

Review journals occasionally devote an entire issue to a single, lengthy article, but usually very lengthy works are published as books. Such books are referred to as "monographs".

Reviews can best be located with the aid of the abstracting and indexing publications and data bases that are discussed later in this chapter. Many chemists and chemical engineers scan their favorite review journals on a regular basis. This approach not only provides them with greater confidence that they have not missed any primary literature relevant to their own specialties but also sometimes provides the serendipity of bringing useful tangential topics to their attention that they might otherwise have overlooked.

Libraries classify monographs for shelf location and in most cases assign subject headings for searching in the library's catalog. The larger academic and research libraries, including many of those in the chemical industry, rely on the guidance of the U.S. Library of Congress for subject headings and classification codes for monographs and other books. Library of Congress subject headings apply broadly to the subject of a book as a whole but give little indication of the detailed contents. The recent introduction of the SUPERINDEX data base is a step in the direction of providing one-stop access to the actual contents of books. SUPERINDEX is an online data base built up from the index entries at the backs of many monographs and reference books in chemistry and other areas of science and technology.

Information Retrieval Services

Information retrieval is the art of finding out what is in the primary and review literature. For an individual chemist or engineer, that may mean finding it for the first time or it may mean pinpointing the location of something vaguely remembered. Information retrieval is facilitated by libraries, librarians, information scientists, and published and computer-readable information retrieval tools. Libraries are organized around concepts of making information and literature retrievable; librarians and information scientists have professional education and experience for facilitating information retrieval. This section focuses on publications and computer-readable data bases that are most useful for information retrieval in chemistry and chemical engineering.

Usually the first references to a primary publication occur in indexes that can be produced without requiring any intellectual content analysis. These are current awareness bulletins, and they will be discussed later. First, however, we will consider those services that provide more in-depth subject analysis. At least some of the following features must characterize the treatment of a document by this kind of service: indication of authorship; an abstract or summary of the work; one or more subject index entries derived from analysis of the entire work or at least from its abstract (not merely from the title); classification of the work into one or more broad subject categories; and other indicators such as language, document type, and the presence of specific kinds of data. All secondary services must provide adequate bibliographic identification (i.e., sufficient information for unambiguously identifying the work and locating a copy), or they are not worth our consideration at all.

Secondary abstracting and indexing publications began appearing near the end of the eighteenth century. Prior to that, many primary journals included abstracts of relevant material from other sources. Even today, a few primary journals regularly feature abstracts of other publications.

Computers are ideal instruments for sorting, indexing, alphabetizing, and keeping track of vast numbers of records—all requisite tasks in the production of abstracting and indexing journals. Once the data were in machine-readable form for the production of the journals, the services realized that they could market an alternative tool for information retrieval: the computerized data bases. Today, nearly all the major abstract and index publications have one or more data base counterparts. Organizations may purchase data bases for in-house processing, but more commonly they contract with search services that allow online searching from remote terminals by multiple users. Data bases will be discussed in more detail later.

Chemical Abstracts

The most important secondary publication for chemists and chemical engineers is *Chemical Abstracts*. Coverage of the chemical literature is truly comprehensive: patents, primary journal articles, review articles, conference papers, and technical reports from all over the world are abstracted and indexed for the weekly issues of *Chemical Abstracts*. More than 12,000 sources are covered. In addition, some monographs and dissertations are announced and indexed, but not abstracted. CASSI, which was mentioned earlier, lists the journals and conference proceedings that are covered.

The document records are arranged in *Chemical Abstracts* in eighty sections based on broad subject classification. Odd-numbered issues include sections from organic chemistry, biochemistry, and chemical history, education, and documentation. Even-numbered issues include sections from inorganic, physical, analytical, and polymer chemistry, applied chemistry, and chemical engineering. Author, keyword subject, and patent indexes appear at the end of each issue.

Chemical Abstracts has in-depth indexes (called volume indexes) for each six-month period. The six-month accumulation of issues is considered a "volume" of *Chemical Abstracts*, and references to *Chemical Abstracts* records are given as the volume and abstract numbers. The volume indexes are not simply cumulations of the indexes at the ends of the individual issues. The differences are most noteworthy for the subject indexes: documents are analyzed in greater depth for the subject volume indexes, and whereas the issue indexes are based on author terminology without vocabulary control, the volume indexes rely, at least in part, on a controlled vocabulary.

Each volume has separate volume indexes for specific chemical substances ("Chemical Substances Indexes") and all other subjects ("General

Subject Indexes"). An "Index Guide", published periodically as part of the volume indexes, serves as a thesaurus to the "General Subject Index". Vocabulary control for the more than seven million substances indexed in the "Chemical Substances Indexes" is based upon an elaborate set of rules.

Policies for indexing and naming compounds have undergone numerous changes since the first issues appeared in 1907. Donaldson and his colleagues (15) explained the significant changes that were made in 1972, which are, with minor modifications, still in effect.

The "Index Guide" outlines the most important *Chemical Abstracts* nomenclature rules and also provides cross-references from some common names to the accepted names. Published and online searching tools have gone a long way toward reducing the need for the information seeker to understand all the complexities of *Chemical Abstracts* nomenclature. Volume index components based on molecular formulas and ring systems often provide relatively painless entries into *Chemical Abstracts* for specific substances or groups of closely related substances.

Chemical Abstracts Service (CAS) assigns a Registry Number to each unique chemical substance that it indexes. A series of articles (16–24) describes the Registry Number system in detail. About 420,000 new compounds per year are added to the CAS Registry system, which contains more than 7 million chemical substances reported in the literature since 1965. Because the CAS Registry Number assignment is done carefully, and with the aid of a very sophisticated computer storage and retrieval system, authors who take care to name substances unambiguously (not necessarily in exact accord with *Chemical Abstracts* rules) or to draw and label structural drawings unambiguously can anticipate that the substances they report will find their proper places in the "Chemical Substance Indexes".

All the registered substances are listed in numerical order in the "Registry Handbook: Number Section", and the companion "Registry Handbook: Common Names" provides an extensive list of common names (and some trade names) with their corresponding Registry Numbers. Most substances, of course, do not have common names and are not listed in the Common Names section. Registry Numbers are provided in square brackets with the entries for all substances in the "Chemical Substance Indexes" and the "Formula Indexes".

Many familiar reference books, including the latest editions of the *Merck Index* and the *Dictionary of Organic Compounds*, include *Chemical Abstracts* Registry Numbers. The *Ring Systems Handbook*, published by Chemical Abstracts Service, is a very useful nomenclature aid for compounds containing ring systems.

For *Chemical Abstracts*, each five-year period is referred to as a "collective period". Prior to 1957, collective periods were ten years long, and were called "decennial periods". Massive indexes that are cumulations of the volume indexes of the preceding five or ten years have been issued at the ends of each of the collective and decennial periods.

The brief description of *Chemical Abstracts* just given touches on some of the most important aspects of its indexing policies, particularly as they apply today. More extensive literature guides (25–32) should be consulted for more details about using *Chemical Abstracts* and other information resources.

Current Abstracts of Chemistry and Index Chemicus

For the organic chemist, *Current Abstracts of Chemistry and Index Chemicus* (CAC&IC) provides selective guidance to slightly more than one hundred important journals. Garfield and his colleagues (33) have shown that more than ninety percent of all new organic compounds that are reported are covered in one hundred to two hundred key journals. Antony and Stevens (34) have compared the coverage of *Chemical Abstracts* with CAC&IC on the basis of a sample of new compounds indexed in CAC&IC, and they found that for both services, indexing practice is very consistent with stated policies.

Chemical Abstracts has a much larger document base than CAC&IC. Although neither service achieves one hundred percent coverage of new organic compounds, *Chemical Abstracts* nearly always provides more substance index entries per document and adds more than twice as many new compounds to its data base than CAC&IC. CAS includes not only newly synthesized substances, but also those substances that have new information reported about them (e.g., a new preparative method, a new reaction, newly reported mechanism studies, chemical and physical properties, methods of detection, or a new use or biological effect). Nonisolated intermediates are indexed by CAC&IC, but are not covered by *Chemical Abstracts* according to its stated policy. CAC&IC is frequently faster than *Chemical Abstracts* in its coverage, generally by several weeks.

Two very valuable features of CAC&IC are the excellent graphics in the records and the fact that compounds reported for the first time are highlighted in the indexes. Molecular formulas illustrate the courses of the most important chemical reactions described in the original articles.

In addition, each record contains an analytical "rectangle" (formerly a "wheel") to help the user tell at a glance which analytical and other techniques were reported. Several specialized indexes are included—for example, one for biological activities and another for instrumental data.

A companion publication, *Current Chemical Reactions*, reports on new reactions and important modifications to established reactions.

Beilstein and Gmelin

Beilstein's *Handbuch der Organischen Chemie* (hereafter referred to as Beilstein) is one of the most valuable reference sources available to the organic chemist. Here we are interested mainly in its function of directing the user to the primary literature, but Beilstein is also a data compilation and a treatise. Gmelin's *Handbuch der Anorganischen Chemie* serves similar functions for inorganic chemistry.

Gmelin and Beilstein are particularly authoritative indexes to chemical literature because information and data are included in the records only after they have been critically evaluated by the experts at the Gmelin and Beilstein Institutes. Critical evaluation is not ordinarily considered the function of a secondary service, and for most secondary publications, including *Chemical Abstracts*, there is no evaluation beyond the decision to cover a particular source or not.

The Hauptwerk (or main work) of Beilstein covers organic chemistry literature through 1909. Subsequent periods are covered by a series of supplements (Erganzungswerken), and the fourth supplement brings coverage through 1959. The fifth supplement, bringing coverage through 1979, has just started appearing. The earlier literature is always interpreted in the light of the most recent information available at the time of publication, so that the fourth supplement contains some references beyond 1959.

To use Beilstein efficiently, it is necessary to understand the structurally based system of arrangement. The Beilstein Institute has recently released a very lucid guide: *How to Use Beilstein*.

The periodic table serves as the basis for the arrangement of Gmelin. In general, each Gmelin "volume" is devoted to a specific element, with compounds going into that volume for the highest numeric component element.

Traditionally, Gmelin and Beilstein were in German. However, the newer parts of Gmelin are in English, as is Beilstein's fifth supplement.

Current Awareness Bulletins

Some publications are designed primarily to alert the chemist to the most recent information that has appeared in print. These publications are current awareness bulletins, structured for rapid reporting, not in-depth analysis. Thus, they complement the secondary publications that were just discussed. The typical current awareness bulletin can be produced by feeding table-of-contents information into a computer and outputting indexes of author names and keywords from article titles.

Chemical Titles is a typical (and important) example. The publisher, Chemical Abstracts Service, reproduces the tables of contents of some seven hundred chemistry journals, and then indexes by author and by keywords from the article titles. To produce the index, the computer has been programmed to permute the words in each title, to center each keyword while not centering certain "stop" words like "and", and to alphabetize the resulting index entries. The result is a keyword-in-context (KWIC) index to the tables of contents.

The Institute for Scientific Information produces a series of current awareness indexes called *Current Contents*. The section devoted to the physical, chemical, and earth sciences is widely circulated among chemical researchers and practitioners. Other sections in life sciences and in engineering, technology, and applied sciences may be preferred by some segments of the chemical community. One of the indexes to each issue of *Current Contents* is based on words from article titles.

Chemists commonly use the current awareness publications in either (or both) of two ways to keep abreast of a subject: they scan the tables of contents that are reproduced from journals that are of particular interest, and they search the keyword title indexes for specific terms. In either case, it is clear that the author's choice of title will have considerable influence on whether his or her article is selected by a potential reader.

Citation Indexes

The authors of scientific documents relate their work to previous publications by citations. Those citations constitute the entry points in a citation index. For example, if author Brown cites an earlier work by author Green, and we know that Green's paper deals with a subject of interest, looking up Green in a citation index leads us to Brown's paper. Thus, citation indexing is a type of subject indexing that is independent of terminology.

Science Citation Index is the most important citation index for chemistry. It is not limited to chemistry, and in fact is often useful for tying chemical topics in with other scientific disciplines such as physics or biology.

Why do authors cite documents? The question has been researched (35), but the answer is complex. Among the many reasons suggested for citing behavior are the following: to acknowledge genuine communication of results or ideas of previous authors, to refute earlier work, to demonstrate that the author has been keeping up with the literature, to buttress an idea or conclusion with the support of an authority, to aid the reader in locating additional pertinent literature, to curry favors, to reward friends, and to insult enemies. Some of these reasons serve the information retrieval function of citation indexes, but others do not. Moravcsik and Murugesan (36) identified forty percent of the references from a sample of high energy physics papers as "perfunctory". In a similar study, Chubin and Moitra (37) classified about twenty percent as perfunctory. These figures suggest that well over half the citations are not perfunctory, and hence predict a relatively high success rate in the use of citation indexes. However, a given paper may discuss many specific topics, and even a highly relevant citation may be to a different topic from the one on which information is sought. For example, an author may cite a paper because of some experimental details relating to the preparation or analysis of a specific compound, whereas the information seeker may be interested in following the progress of ideas that were introduced in that paper.

Heavily cited papers (or "classics") have special characteristics. Small (38) suggests that when an author cites a heavily cited paper, that author is establishing a link between his or her paper and an established part of the current paradigm of that field. The citation is, in effect, a "symbol" for that part of the paradigm. Some heavily cited papers in chemistry introduced or significantly improved laboratory methods that are widely used; others contain ideas or concepts that have assumed key significance in the progress of the field. Near the beginning of every issue of *Current Contents* there is a brief report on a heavily cited document by one of the authors of that document. These reports provide a revealing record of scientific progress and the role that scientific literature has played in that progress.

Other Abstracts and Indexes

Numerous specialized chemical indexes are published, such as *Analytical Abstracts* and *Theoretical Chemical Engineering Abstracts*. Some are intended primarily for current awareness in a particular area. Others are useful at times for retrospective searches by providing greater focus (and hence ease of use) than the more general chemistry indexes. There are a few specialized chemistry indexes, such as William Theilheimer's *Synthetic Methods of Organic Chemistry*, which present unique indexing approaches for their specialized areas, and are so thorough in their coverage that they may be (in some circumstances) the best first choice for a thorough retrospective search. Theilheimer's work is an annual index to the methods of synthetic organic chemistry, with emphasis on bonds formed and broken in the reactions.

Also available are general indexes to science and to broad areas of science that are on the disciplinary borders of chemistry. The H. W. Wilson Company publishes a series of indexes to the high-circulation English-language magazines that are apt to be found in public and small college libraries as well as the larger research facilities. Two of these, *Applied Science and Technology Index* and *General Science Index*, include many items of chemical interest. The former may be particularly valuable for finding a few good articles on applications of chemical technology that are of general interest. These two Wilson indexes, as well as the *Readers' Guide to Periodical Literature*, are valuable for nonspecialists (such as high school students) interested in obtaining some limited, nonspecialist information on a chemical topic.

Referativnyi Zhurnal, produced in the Soviet Union, and *Bulletin Signaletique*, produced in France, are both large, multidisciplinary indexes, with abstracts, that attempt comprehensive coverage of the world's scientific literature. Both have large sections that are relevant to chemistry. However, for most chemistry information needs they need not be consulted, because excellent English-language sources are available. *Chemisches Zentralblatt*, the German index that ceased publication in 1969, was probably the most important rival to *Chemical Abstracts*, and is especially useful for pre-1935 literature.

Chemists and chemical engineers sometimes need to consult indexes and abstracts in other subject areas, especially when interdisciplinary research is involved. *Engineering Index* is an excellent resource for all areas of engineering, including chemical engineering. *Engineering Index* subject indexing is relatively shallow compared to that of *Chemical Abstracts*, and patents are not covered. On the other hand, the broad coverage of engineering topics in *Engineering Index* can be an asset that leads to the retrieval of relevant items in a chemical engineering search that were missed in *Chemical Abstracts*.

Metals Abstracts covers the physics, chemistry, and technology of metals and their alloys. Although *Metals Abstracts* overlaps with *Chemical Abstracts*, a comprehensive metals-related search requires both sources. So many other major literature indexes are sometimes useful in chemistry that they cannot all be named here. The following titles are suggestive of the scope of possibilities: *Biological Abstracts*, *Physics Abstracts*, *Mathematical Reviews*, *Bibliography and Index of Geology*, *Index Medicus*, *Current Index to Journals in Education*, and *Computer and Control Abstracts*.

An extensive listing of abstracts and indexes in science may be found in the guide by Owen and Hanchey (*39*). An even more extensive directory to abstracting and indexing services in general (*40*) was recently published by Gale Research.

Online Data Base Searching

Chemical Abstracts Service and most other secondary services have computerized their publishing operations. The data that go into the secondary publications are in computer-readable form, and thus already captured for searching by computer. Originally, from an economic point of view, the secondary services considered their computer-searchable products as byproducts of the publication stream. However, for some services, the computer data bases are now more important as sources of revenue than the publications.

Most data base searching is done today on remote terminals that are connected via telecommunication lines to the computers that hold the data bases; because the searching is done in real time, it is referred to as online searching. Chemical Abstracts Service has established its own online search service, called CAS ONLINE. In addition, a number of search services subscribe to *Chemical Abstracts* and other data bases. The search services each have established a "command language" for searching, in which the searcher sits down at a terminal, logs onto the computer, selects an appropriate data base, and then issues a series of commands to generate a bibliography on a subject of interest. Alternatively, a search may be conducted for all the journal articles by a particular author, or all the patents assigned to a particular company, or, in general, for whatever kinds of literature references that might have been sought in published abstracts and indexes.

Search terms are not limited to those that are in the printed indexes, and they may be combined with one another with Boolean (AND, OR, NOT) or other operators. For example "50-00-0 (the Registry Number for formaldehyde) AND nmr" limits a search to documents that treat both formaldehyde and nuclear magnetic resonance, whereas "50-00-0 OR 75-07-0" (the Registry Number for acetaldehyde) will retrieve documents that treat either (or both) of the substances represented by the Registry Numbers. Command languages of each system have unique features that can make a difference in the cost of a search and the quality of the output.

In addition to searching directly with CAS ONLINE, the most important search services for chemists in the United States (with their command languages indicated in parentheses) are Bibliographic Retrieval Services (BRS), Lockheed Dialog Information Services (DIALOG), Pergamon Infoline Ltd. (INFOLINE), Questel, Inc. (QUESTEL), and System Development Corporation (ORBIT).

All of these search services subscribe to the principal data base from Chemical Abstracts Service, CA SEARCH. In addition, each service subscribes to numerous other data bases. The mix of data bases varies from one service to another, and some data bases are exclusively available from one service. For brevity, I will use the command language (e.g., BRS or DIALOG) to refer to either the language or the search service.

CA SEARCH and CA FILE The CA SEARCH data base closely parallels the *Chemical Abstracts* publication. Each document record on CA SEARCH contains all the bibliographic information that is found in *Chemical Abstracts* (but not the abstract), plus all the index entries that apply to that document and some additional codes, such as one indicating the language of the document. The chemical substance information is in the form of Registry Numbers rather than fully spelled-out names.

Each search service handles the CA SEARCH file differently (41), but in general, a searcher may combine author names, Registry Numbers, title words, general subject headings, words from the modifications to the general subjects, issue index keywords, language codes, patent data, etc., using operators to tailor a very specific search.

CA SEARCH covers from 1967 to the present, but not all of the services provide complete coverage. Because of the size of the file, those services that do provide complete coverage of the available time period have split the file into several time range parts, each of which must be searched separately. Search-save features facilitate setting up the search once, and then running it sequentially against each of the different parts.

The CAS ONLINE data base, CA FILE, is similar to CA SEARCH, with two important differences: Abstracts are displayable (but not searchable), and all of the records are in a single file covering 1967 to the present. For searchers who are familiar with several systems, the decision to use one rather than another for a specific search is based on weighing a number of factors: the facility with which the command language can achieve the desired results, the print or display capabilities, specific factors related to how the records are handled for the data base of interest, the availability of other relevant data bases, and the probable cost.

Until very recently, the *Chemical Abstracts* Registry Number system applied only to substances indexed since 1965. All the well-known, older substances are, of course, included, as are new substances reported for the first time since 1965. Now Chemical Abstracts Service is engaged in pre-1965 substance registration. This task is part of a program to extend online coverage of *Chemical Abstracts* backwards in time. The CA FILE data base corresponds to *Chemical Abstracts* from 1967 to the present, and another file, CAOLD, has been introduced on CAS ONLINE to cover pre-1967. CAOLD, which is a very small file at present, provides references to records in the printed version of *Chemical Abstracts* but does not contain any bibliographic information or abstracts.

Substructure Searching   Much of chemistry, and the technology based on chemistry, can be described in terms of molecular structures and the changes that take place in them. Thus, molecular structural representations are of fundamental importance in chemical information retrieval. The chemist often has a specific substance in mind when he or she conducts a literature search. Traditional molecular formula and name indexes have usually been proven adequate for that kind of information need. Often, however, the search is for part of a structure (a substructure) rather than for an exact structure. Interest in substructures stems from the fact that structural features correlate with chemical properties and behavior.

Chemical nomenclature has not been adequate for indexes designed for substructure searching. Names reflect some features but overlook others. By permuting a name and thereby bringing various parts of the name to the beginning, it is possible to create several index entries for the same substance, such that each entry emphasizes a different structural feature. But permuting does not adequately reflect all of the structural features, and has the disadvantage of significantly increasing the length of an index. Ingenious linear notation systems have been designed to represent three-dimensional structures in a single line of alphabetical and numerical symbols. The Wiswesser notation is the most widely used linear notation, and permuted Wiswesser representations comprise a number of literature and reference book indexes. Although superior to nomenclature for substructure searching, linear notations suffer to some degree from the same limitations and are unfamiliar to most chemists.

The *Chemical Abstracts* Registry system, which assigns a unique Registry Number to each unique structure reported in the literature, is based on connection tables. The table lists each atom with its bonds (i.e., connections) to every other atom. The connection tables in computer-readable form can be searched in whole or in part, the latter constituting a substructure search.

CAS ONLINE and QUESTEL have made the *Chemical Abstracts* Registry system available for online searching. With either of these systems, it is possible to define a substructure in text form, by describing the atoms of the substructure and the bonds that link them. It is also possible to draw the substructure on a graphics terminal. Whichever approach is used, the system searches for all those substances that contain the substructure, and includes provisions for obtaining literature references for those substances.

An alternative approach to substructure searching is provided by ORBIT and DIALOG. Both ORBIT and DIALOG have chemical dictionary files containing chemical nomenclature, synonyms, molecular formulas, and ring data for all the substances that have *Chemical Abstracts* Registry Numbers. A similar dictionary file is also available on the NLM search system (from the U.S. National Library of Medicine), which is used primarily by people in the health and biomedical fields. The usual procedure is to search the dictionary files using names or parts of names and other search parameters to generate a list of Registry Numbers, and then to use those Registry Numbers in a bibliographic file to obtain literature references. Recently, a chemical dictionary file was also added to CAS ONLINE.

Searching on Other Data Bases   One of the advantages of online searching is that an enormous number of different data bases are available, so that more than one resource may be brought to bear on a subject at a single session at the terminal. For example, the January 1985 Dialog Database Catalog lists more than 200 data bases available on DIALOG. Some of these data bases contain little or no information of chemical interest, but many are, to varying degrees, of use to the chemical searcher. For example, INSPEC, which corresponds to several published indexes including *Physics Abstracts* and *Computer and Control Abstracts*, is useful for physical chemistry and applications of computers in chemistry. An INSPEC search may retrieve items that are missed on CA SEARCH for several reasons: The document may not be covered by the CA SEARCH data base; INSPEC indexing is different from CA SEARCH indexing; the record for a document may be added to INSPEC before the corresponding record is added to CA SEARCH; and words from abstracts are searchable on INSPEC but not on CA SEARCH. Of course, CA SEARCH may retrieve documents that are missed on INSPEC. For interdisciplinary subjects, especially, it is often important to use more than one independent source in a search.

Some data bases are particularly rich in chemical information. Only a few can be described here. APILIT and APIPAT, produced by the American Petroleum Institute and searchable on ORBIT, cover the literature and patents, respectively, dealing with oil refining, petrochemicals, and other subjects of interest to the petroleum industry. Indexing is in depth and based on a carefully developed vocabulary. The API data bases have very sophisticated features that enhance retrieval effectiveness. For example, all subject terms are automatically posted to broader terms, eliminating the need for the searcher to think of and type in all the more specific terms when searching on a broad concept. The API vocabulary includes some broad chemical structure features that often offer an advantage over the more detailed substance-specific indexing in CA SEARCH. On the other hand, the API data bases do not offer the capability of searching for every specific compound of interest (which, in principle, can be done on CA SEARCH), and, indeed, the document base for the API data bases is much more limited than for CA SEARCH.

The distinctions between the API data bases and CA SEARCH are belabored here to emphasize the value of multiple, independent resources for information retrieval. Even for searches that are not interdisciplinary, more than one resource should be used. Or, if only one data base can be used, the advantages of several should be weighed beforehand.

RAPRA, on INFOLINE, which covers the scientific and technical literature relevant to plastics and rubbers, and METADEX, on both ORBIT and DIALOG, which corresponds to *Metals Abstracts*, are examples of two other data bases with a high relevance to chemistry. The most important multidisciplinary data base for chemists is SCISEARCH, on DIALOG, which corresponds to *Science Citation Index*. Citation searching can be done on SCISEARCH, as well as searching by author, title word, or other parameter. Subject index terms from a controlled vocabulary are not included on SCISEARCH, so its principal advantages are its multidisciplinary coverage and citation searching capabilities. INDEX CHEMICUS ONLINE, a data base corresponding to *Current Abstracts of Chemistry and Index Chemicus*, is available on QUESTEL.

For interdisciplinary topics, it is often mandatory to include data bases such as BIOSIS (searchable on BRS, DIALOG, or ORBIT) or GEOREF or ENVIROLINE (searchable on DIALOG or ORBIT). The NTIS data base (for technical reports), which is searchable on BRS, DIALOG, or ORBIT, is particularly useful for applied topics for which there is government-sponsored research. SUPERINDEX, searchable exclusively on BRS, has already been mentioned as a valuable reference tool for searching back-of-the-book indexes.

Patent Data Bases   The *Chemical Abstracts* data bases, CA SEARCH and CA FILE, cover U.S. and foreign chemical patents and include access via title terms, index terms, Registry Numbers, etc., just as they do for other kinds of documents. In addition, some special "fields" or search access points are used for patents, such as the patent number. Some of the other data bases that were just discussed, such as RAPRA, also cover patents, but many of them do not.

For thorough patent searching, it is important to search one or more of the data bases that are designed specifically for patent searches. These specialized data bases are successful for a number of reasons, but two aspects of patent searching are particularly problematical and in part explain the proliferation of patent-related data bases: patent families and Markush structures.

A patent family is the collection of all the patents that issue from the same application. Usually, the different family members are for the same invention in different countries. The INPADOC data base on INFOLINE, prepared by the International Patent Documentation Center, is the most comprehensive in terms of countries covered. A separate data base, INPANEW (also on INFOLINE), covers the same patent offices as INFOLINE, but includes only very current data. Records are removed from INPANEW after a few weeks. Thus INPANEW is a data base designed for current awareness.

INPI-3, on QUESTEL, may also be used for patent family searching. INPI-3 covers fewer countries than INPADOC, but is cheaper. WPI and WPIL (discussed in detail later) are used often for family searching: they are even less expensive than INPI-3, but cover even fewer countries. However, they do cover all of the countries that are usually considered most important for the typical chemical patent search.

WPI and WPIL are complex data bases that afford many search capabilities in addition to the ability to search for family members. Both of these data bases are on ORBIT, DIALOG, and QUESTEL. WPI is for pre-1981 patents, and WPIL for 1981 to present. For the sake of brevity, I will refer to both of these data bases collectively as WPI from here on.

Derwent Publications Ltd., the producer of WPI, is the publisher of a rather comprehensive array of indexes and current awareness bulletins to patents issued all over the world. A "full" subscription to Derwent services (which does not necessarily include all the publications) is very expensive, but the quality is high and many companies need the thorough patent coverage that Derwent provides.

The many subject-related access points on WPI reflect the detailed subject analysis. Chemistry is emphasized. Chemical structures are tackled on WPI mainly with codes. These codes are particularly valuable for retrieving patents in which chemical substances are defined by Markush structures.

A Markush structure is a chemical structure in which some atoms and their connections are specified, but others are allowed to vary in some way. Markush structures are allowed in patents to protect the invention for sets of related compounds without having to require the inventor to test each and every possible example. Kaback (42) discusses an example he found of a Markush structure that could generate more actual compounds than the total number of substances registered by Chemical Abstracts Service. Chemical Abstracts Service does not assign Registry Numbers to Markush structures but does assign Registry Numbers to those specific compounds for which there are actual experimental data in a patent or which are identified in the claims (and hence may constitute specific examples of a Markush structure).

The combination of chemical codes and ring codes in a WPI search constitutes a substructure search for the retrieval of patents in which Markush structures as well as specific substances are described. This type of searching is particularly valuable for pharmaceuticals and agricultural chemicals. A separate set of codes is applicable to polymers. Only full subscribers to Derwent services have access to all of the codes.

IFI/Plenum issues data bases for U.S. patents that are coded for molecular structures. A series of these data bases is searchable on DIALOG under the name CLAIMS. Although CLAIMS is limited to U.S. patents, it goes back to 1950 (at least for chemical patents), and is the only patent-related data base with sophisticated subject access to go back that far in time.

Many patent searches have legal ramifications, or may determine whether an expensive research project is undertaken. For such searches, it is essential not to miss any relevant patents. Investigators have shown (43) that when high recall is essential, multiple data base searching is a necessity. For example, searches on APITAT, CA SEARCH, CLAIMS, and WPI might be needed to locate all of the patents that are relevant to a particular petrochemical invention. High-recall searches usually have low precision: to get all the relevant records, a lot of irrelevant ones have to be retrieved as well.

A few other patent-related data bases deserve brief mention. USPA, USP77, and UPS70 (on ORBIT and collectively referred to as USPATENTS) cover U.S. patents back to 1970. The records are very long and contain large segments of the patents covered, so that these data bases approach the nature of full-text data bases. LEXPAT, available from Mead Data Central, is a full-text patent data base that will be discussed later. PATSEARCH on INFOLINE covers U.S. patents issued since 1971, with records that are not as lengthy as those on the USPATENTS data bases. Videodiscs on which drawings from patents are stored may be used in conjunction with PATSEARCH.

The PATS data base on BRS has relatively brief records (with abstracts) of U.S. patents issued since 1975. It is perhaps the least expensive U.S. patents data bases to search, and is useful for inventor or assignee searches, and for subject searches for which high recall is not a requirement.

PATS, USPATENTS, and PATSEARCH also allow for citation searching. CLAIMS CITATION on DIALOG is a special IFI/Plenum data base for patent citation searching that tends to be more expensive than the others but covers a greater time span.

Full-Text Searching   One of the latest developments in commercial information retrieval is the release of data bases that contain the full texts of journal articles, patents, or other documents. With these data bases, it is possible to search for words or combinations of words within any part of the text and to display all or part of the entire document. Often, full-text data bases produce results that are difficult to achieve with data bases containing document surrogates, especially for searches that require the use of jargon or newly developed terminology. The depth of indexing clearly is superior on full-text data bases. For certain information needs, there is also an advantage in being able to pick out and display only the most pertinent parts of a document without having to look at it all. The full-text data bases are perhaps forerunners of electronic journal systems (44) where documents are prepared, submitted, refereed, edited, retrieved, and read online, and then selectively chosen for filing in a personal online data base.

Nonetheless, full-text searching presents some disadvantages. Considerable care must be exercised by the searcher so that a lot of irrelevant material is not retrieved. None of the full-text data bases available now in chemistry carry with them any of the sophisticated coding or indexing that characterize most of the other data bases discussed in this chapter. As a result, it is easy to retrieve irrelevant documents, and in some cases it is all too easy to miss some that are relevant. Full-text data bases may, in the future, by including additional searchable fields containing index terms, chemical codes, etc., reduce the retrieval problems presently associated with them. At present, full-text data bases provide only very limited coverage of the total amount of literature needed for many searches.

The American Chemical Society has introduced ACS JOURNALS ONLINE, which contains the full texts (without tables or figures) of its nineteen primary journals as the data base CFTX (on BRS). Coverage of most of the journals is from 1980 to the present. BRS software includes several features for focusing both search and display on the most relevant parts of the articles. Any part of a journal article can be searched and displayed. Every word is searchable. Of particular interest is the capability to search for very specific kinds of information such as spectroscopic techniques, molecular formulas, thermodynamic data, toxicity data, biological data, and chemical names. The data base is updated every two weeks. More than 50,000 complete journal articles are included. ACS has also announced plans to put this data base on STN International .

Mead Data Central has concentrated on the full-text data base field, with emphasis on legal and news-related documents, including newspapers and magazines. Of particular interest here is LEXPAT, which has the full texts of U.S. patents issued since 1975. Full texts of magazines, including a few of interest in chemical marketing such as *Chemical Week*, are included in NEXIS.

Numerical data bases are numerous, and will receive only brief mention here. Numerical data bases containing physical and chemical property data may be of particular interest to chemists. Many data bases of economic statistics are of interest in the chemical and related industries. A few numerical data bases (mainly economic) are available on DIALOG, ORBIT, and other services already discussed.

The NIH-EPA Chemical Information System specializes in numerical data bases in chemistry and toxicology and provides access to a few bibliographic data bases as well. The NIH-EPA Chemical Information System includes its own structure-based file of chemicals (SANSS) on which substructure searches may be performed. SANSS contains about 350,000 substances, which is far fewer than the over seven million substances in the files based on the *Chemical Abstracts* Registry System. The advantage of SANSS is that it can be related directly to the other NIH-EPA Chemical Information System components.

Cuadra Associates publishes a rather comprehensive directory of online numerical and bibliographic data bases four times a year (45).

Selective Dissemination of Information

Selective dissemination of information (SDI) is the term information people use to refer to those activities that help researchers stay up to date with the literature on a specific topic. For example, a library might route the issues of *Current Contents* to the members of a department. It is possible to save a search strategy used for an online data base search and then run it periodically against updated parts of the data base. It is also possible on most search systems to issue a series of commands to have the strategy run automatically against each new update of one or more data bases and to have printouts made on a regular basis. For convenience, I will call that kind of stored strategy an "SDI profile".

SDI profiles may be submitted directly to some major data base and index producers, including Chemical Abstracts Service and the Institute for Scientific Information. Alternatively, a profile may be set up on DIALOG, ORBIT, or some other search service, and run automatically against selected data bases. The advantage of either of these approaches is that the SDI can be tailored to exact interests and modified when needed to reflect a change in interests.

However, if a number of people share very similar interests, they may save money by sharing SDIs. Chemical Abstracts Service issues a series of publications, called CA SELECTS, that provide SDI outputs on topics of common interest to a number of people. The American Petroleum Institute's *API Bulletins* serve a similar function.

Acknowledgments

I wish to acknowledge the valuable advice that I received from William W. West of Chevron Research Company in preparing this chapter, especially regarding patent searching. I also acknowledge the helpful responses I received to my request for advice from W. V. Metanomski of Chemical Abstracts Service, I. Shaw of Derwent Publications, Ltd., Irving Zarember of the American Petroleum Institute, and K. C. Buschbeck of the Gmelin Institute.

Literature Cited

1. Garvey, W. D.; Lin, N.; Nelson, C. E.; Tomita, K. *Inf. Storage Retr.* **1972**, *8*, 111-122.
2. Garvey, W. D.; Tomita, K.; Woolf, P. *Inf. Storage Retr.* **1974**, *10*, 115-131.
3. Crane, Diana *Invisible Colleges: Diffusion of Knowledge in Scientific Communities;* University of Chicago: Chicago, 1972.
4. Kuhn, Thomas S. *The Structure of Scientific Revolutions;* University of Chicago: Chicago, 1970, 2nd ed.
5. *Comprehensive Chemical Kinetics;* Bamford, C. H.; Tipper, C. H. M., Eds. Elsevier: Amsterdam, 1969.
6. *Comprehensive Inorganic Chemistry;* Bailar, J. C., Jr.; Emelius, H. J.; Nyholm, Sir Ronald; Trotman-Dickenson, A. F., Eds. Pergamon: Oxford, 1973.
7. Price, Derek J. de Sola *Little Science, Big Science;* Columbia University Press: New York, 1963.
8. *Statistical Indicators of Scientific and Technical Communication (1960-1980). Volume II: A Research Report;* Prepared by D. W. King and others; National Technical Information Service: Springfield, VA, 1976. (PB-254060). pp 117-121.
9. Grubb, Philip W. *Patents for Chemists;* Clarendon: Oxford, 1982.
10. Olby, Robert *The Path to the Double Helix;* University of Washington: Seattle, 1974.
11. Garvey, W. D.; Lin, N.; Nelson, C. E.; Tomita, K. *The Role of the National Meeting in Scientific and Technical Communication;* Johns Hopkins University: Baltimore, 1970. Also published by the U.S. National Technical Information Service (PB-202367).
12. Garvey, W. D.; Lin, N.; Nelson, C. E.; Tomita, K. *Inf. Storage Retr.* **1972**, *8*, 159-169.
13. Bottle, R. T. *J. Doc.* **1973**, *29*, 281-294.
14. Woodward, A. M. *Problems and Possible Investigations in the Study of the Role of Reviews in Information Transfer in Science;* Aslib: London, 1975. (British Library R&D Report No. 5234).
15. Donaldson, N.; Powell, W. H.; Rowlett, R. J., Jr.; White, R. W.; Yorka, K. V. *J. Chem. Doc.* **1974**, *14*, 3-15.
16. Dittmar, P. G.; Stobaugh, R. E.; Watson, C. E. *J. Chem. Inf. Comput. Sci.* **1976**, *16*, 111-121.
17. Freeland, R. G.; Funk, S. A.; O'Korn, L. J.; Wilson, G. A. *J. Chem. Inf. Comput. Sci.* **1979**, *19*, 94-98.
18. Blackwood, J. E.; Elliott, P. M.; Stobaugh, R. E.; Watson, C. E. *J. Chem. Inf. Comput. Sci.* **1977**, *17*, 3-8.
19. Vander Stouw, G. G.; Gustafson, C.; Rule, J. D.; Watson, C. E. *J. Chem. Inf. Comput. Sci.* **1976**, *16*, 213-218.
20. Zamora, A.; Dayton, D. L. *J. Chem. Inf. Comput. Sci.* **1976**, *16*, 219-222.
21. Stobaugh, R. E. *J. Chem. Inf. Comput. Sci.* **1980**, *20*, 76-82.
22. Mockus, J.; Stobaugh, R. E. *J. Chem. Inf. Comput. Sci.* **1980**, *20*, 18-22.

23. Moosemiller, J. P.; Ryan, A. W.; Stobaugh, R. E. *J. Chem. Inf. Comput. Sci.* **1980**, *20*, 83-88.

24. Ryan, A. W.; Stobaugh, R. E. *J. Chem. Inf. Comput. Sci.* **1982**, *22*, 22-28.

25. Antony, Arthur *Guide to Basic Information Sources in Chemistry*; Wiley: New York, 1979.

26. Crane, E. J.; Patterson, Austin M.; Marr, Eleanor B. *A Guide to the Literature of Chemistry*, 2nd ed.; Wiley: New York, 1957.

27. *Information Retrieval in Chemistry and Chemical Patent Law*; Grayson, Martin, Ed.; Wiley: New York, 1983. (Articles reprinted from the *Kirk-Othmer Encyclopedia of Chemical Technology*, 3rd ed.)

28. Maizell, R. E. *How to Find Chemical Information: A Guide for Practicing Chemists, Teachers, and Students*; Wiley: New York, 1979.

29. Mellon, M. G. *Chemical Publications: Their Nature and Use*, 5th ed.; McGraw-Hill: New York, 1982.

30. Skolnik, H. *The Literature Matrix of Chemistry*; Wiley: New York, 1982.

31. *Use of Chemical Literature*, 3rd ed.; Bottle, R. T., Ed.; Butterworths: London, 1979.

32. Wolman, Y. *Chemical Information: A Practical Guide for Utilization*; Wiley: Chichester, England, 1983.

33. Garfield, E.; Revesz, G. S.; Batzig, J. H. *Nature* **1973**, *242*, 307-309.

34. Antony, Arthur; Stevens, Janet *J. Chem. Inf. Comput. Sci.* **1980**, *20*, 101-105.

35. Kaplan, Norman *Am. Doc.* **1965**, *16*, 179-184.

36. Moravcsik, Michael J.; Murugesan, P. *Soc. Stud. Sci.* **1975**, *5*, 86-92.

37. Chubin, Daryl E.; Moitra, Soumyo D. *Soc. Stud. Sci.* **1975**, *5*, 423-441.

38. Small, Henry G. *Soc. Stud. Sci.* **1978**, *8*, 327-340.

39. Owen, D. B.; Hanchey, M. M. *Indexes and Abstracts in Science and Technology: A Descriptive Guide*; Scarecrow: Metuchen, NJ, 1974.

40. *Abstracting and Indexing Services Directory*; Schmittroth, John, Jr., Ed.; Gale Research: Detroit, 1982-1983.

41. Antony, Arthur *Database* **1983**, *6(3)*, 76-80.

42. Kaback, Stuart M. *J. Chem. Inf. Comput. Sci.* **1980**, *20*, 2-6.

43. Kaback, Stuart M. In *Online '82 Conference Proceedings*; Online, Inc.: Weston, CT, 1982; pp 157-159, *Online Patent Searching—The Realities.*

44. Lancaster, R. W. *Libraries and Librarians in an Age of Electronics*; Information Resources: Arlington, VA, 1982; pp 78-87.

45. *Directory of Online Databases*; Cuadra Associates: Santa Monica, CA, quarterly.

7
Chapter

Making Effective Oral Presentations

LARRY VENABLE

O ther than the technical knowledge you have in your field, perhaps no other knowledge or skill is more important than your ability to make presentations. In fact, your ability to communicate may be the most important factor in making you stand out among your peers, who probably have the same education and training as you do.

Whenever you are asked to give a presentation, the first thought you should have is "opportunity". Each time you give a presentation, you have the opportunity to accomplish two objectives: First, you can present the results of your research and your point of view. All the work, knowledge, and effort you have put into your profession are certainly important, but your advantage comes when you finally have the occasion to present your work to groups of people. The work of the greatest scientist in the world is of no value if he or she cannot share information with others in a useful format.

Second, and perhaps even more important, you have the opportunity to make a positive impression. Every time you give a presentation, you have the opportunity to convince people that you have skills beyond the ordinary or that your research will yield important and significant results. To be very blunt, your presentation skills may be a critical factor in your future success and your income.

I cannot emphasize this point strongly enough. For twenty years, I worked at a company well-known for its outstanding technical accomplishments. Almost all of the top executives had technical backgrounds and educations, but their rise to the top occurred because they also had the ability to communicate and to get things done.

As important as this skill is, it is also an area that is greatly feared by a surprisingly large number of people. In a survey conducted for the *Book of*

0917–0/86/0185$06.00/0   © 1986 American Chemical Society

Lists, speaking before a group of people was rated the number one fear by the people of the United States. It ranked ahead of fear of heights, financial problems, deep water, sickness, flying, and death. Why do so many people fear this fundamental segment of business? There are many reasons, among which are the following:

1. Lack of confidence. Speaking before a group demands specific skills. Yet, the odds are great that in your academic years you did not spend much time or effort studying the art of making presentations. Most likely, if you did take one speech course, you treated it as a requirement to get out of the way as quickly as possible. Because you have had very little training, you may lack self-confidence in this area.

2. Fear of failure. Lacking the specific skills of making presentations and realizing that you are being evaluated when you speak, you tend to have an inordinate amount of concern about how you look and how you are doing. The fear of failing in the situation creates panic.

3. Fear of getting up in front of a group. Many people have told me that, during their elementary school days, if they were called upon to get up and recite in front of a class, most of the other kids would snicker and giggle, and it was embarrassing to be singled out. The "losers" got called on, and the "winners" didn't. These experiences have left them with a fear of speaking before a group.

4. Feeling unnatural. Many people believe that getting up in front of people is not natural. That belief surprises me because talking is one of the most natural things that we do: We talk to people all the time. We talk to our friends, our co-workers, our families. Yet for some reason, many people feel that when they get up in front of an audience, everything changes. They feel unnatural and uncomfortable because they have been taught all the wrong tools, all the wrong props, and all the wrong techniques, and have been forced to be unnatural. One of the central themes throughout this chapter will be that you should be yourself and be natural, and I will stress all of the tools and techniques that will help you to be so.

When you are giving a presentation, you are putting four parts of yourself into it: knowledge, skills, attitude, and energy.

Your knowledge is obviously very important. Your experience and your preparation are certainly critical to the success of the presentation, but not necessarily any more important than the skills of communicating the information. But even the skills of communication are not necessarily any more important than your attitude, positive or negative. If you go into a presentation assuming that it is not going to go very well, it probably will not go well. If you have a positive attitude about it, you will do much better. A positive attitude converts into energy. If you are convinced that your idea is valid and that your audience should adopt your proposal, you will convey that conviction and that enthusiasm in your voice; that may be just as important as your technical expertise.

Planning the Presentation

You should start putting your presentation together by answering several questions:

What is the title and what do I want to accomplish?
Who will be there?
Why should they listen to me?
How should I organize my presentation?
How much time do I have?
Shall I write out a script or make an outline?
Where will it be and what props or tools do I need?

Title and Objective

You must have a clear reason for giving a presentation, and you must keep that reason in focus. However, you must also keep in mind that your title and your objective may not be identical. The title of your talk may be a report on the status of a project, but your objective might be to stimulate a discussion that will broaden your research or interpret your results in a way you had not considered, or to convince management to give you more time or more money.

Audience

Without question, the most important part of your preparation is your audience analysis. I cannot emphasize this point strongly enough. Communication is not what you say to people, but rather, what they hear. Communication is not just your intent, it is also the listener's perception. What you said does not matter as much as what the audience heard. How did they interpret it? Did it meet their objectives?

Several factors are involved in an audience analysis:

1. The role of the group. Are these people fellow scientists, management, students, subordinates, customers, salespeople?

2. Experience. What experience do they have in your field?

3. Interest. What are they interested in? For example, a top executive is usually interested in the bottom line. Would a factory worker be interested in the bottom line? Probably not. Each group has its own interest at heart.

4. Need to know. As a technical person, you have a tendency to have one major fault when you present to nontechnical people or to technical people in different disciplines: That tendency is to give too much information. Although you are very comfortable in your field and it is easy for you to talk about topics that you understand so well, a nontechnical or

nonexpert audience is not at that comfort level. You must decide what they need to know to accomplish your objective as well as their objective.

5. Their objective. Why are they here today? What do they want to accomplish?

6. Language level. What terms, jargon, phrases, and technical language can you use safely so that your audience can understand your meaning and intent?

7. Attitude. What is the audience's attitude toward you or your idea or your proposal? Your audience could be respectful and interested to hear what you have to say, they could be neutral or open-minded about hearing your presentation, or they could be in a defensive, antagonistic frame of mind.

8. Will there be people in the audience with special needs, such as hearing or sight problems? It might be possible to accommodate them to some extent, but you would need to know in advance that they will be there. Presumably they need the information as much as everyone else, or they would not be attending. I know someone who gave an entire presentation based on visual aids and found out later that a blind woman was in the audience.

Organization

Most of us have been taught that a good presentation has an introduction, a body, and a conclusion. Although this method of organizing a presentation is accurate, unfortunately it is very incomplete. More important questions are "What goes into an introduction? How do I organize the body?" The following are ten ways to organize a presentation, all of which have an introduction, a body, and a conclusion.

1. Sequential. You could organize your presentation by explaining a series of events or steps in a process such as an assembly line process or the step-by-step development of a new compound.

2. Categorical. You could divide your talk into three or four major categories.

3. Spatial. This technique involves orienting the audience to the entire object and then explaining each individual part as it relates to the whole.

4. Contrast and comparison. You could explain a new procedure in your program, a new project, or a new product by showing the similarities and the differences between the new and the previous one.

5. Objective, status, schedule. At the R&D laboratory of a major manufacturing firm, each research technician is asked to participate in the Friday morning lab review. Each participant is given five minutes to give an oral presentation following this pattern. He or she states the objective of the research, briefly explains its current status, and projects the schedule for the next several weeks or months.

6. Ideal versus reality. A new product could be explained this way: "Ideally, we would like to have all of these features, but because of budget and time restrictions, we can only do the following." In other words, explain what would be perfect and what is actually obtainable.

7. Fact, function, benefit. This approach is often used to explain products to customers. The fact describes what the product or idea is, the function refers to what it does, and the benefit concerns what it means to the potential user in terms of time savings, cost savings, productivity, etc.

8. Problem-solution method. In this method, you state the purpose of your presentation; then you describe the problem, including its historical perspective. Discuss the major constraints, such as legal, budgetary, time, or design. Describe each alternative, and give pros and cons of each. Finally, state your recommendation.

9. Inquiry method. In this approach, you start by recognizing the problem and giving a precise description of its nature and extent. Then, you present objectives, goals, and possible solutions. The next step is to evaluate the solutions. Finally, give your recommendation.

10. The process of elimination. This approach means eliminating several approaches to an idea or solutions to a problem by explaining the disadvantages of each until you reach the one you want to promote. Your proposal usually sounds better when presented in this fashion, because it shows that you have considered other alternatives.

By having an organizational pattern in mind before you start putting your talk together, three things will happen: Your preparation will be easier, you will remind yourself to include key points, and you will sound more logical and more believable.

Time

You must clarify how much time you are to be given for your presentation. Then, do not plan to use every available minute. For example, if you are told that you have exactly fifteen minutes, and you practice a presentation and time it to exactly fifteen minutes, when you give the presentation, you will never finish in time. Because of questions, and because as you are presenting you sometimes realize the inadequacy of certain words and tend to explain, your presentation always will expand from your original expectations. So, plan a somewhat shorter presentation than your allotted time.

Method: Using a Script, an Outline, Index Cards, or Visual Aids

In the seminars that I conduct, I have asked thousands of people the question, "Have you ever attended a meeting or a convention where a paper was read?" Of the people who answer yes, the overwhelming majority of them have said it was boring, uninteresting, dull, insulting, and sleep-provoking. Typical statements were "I just couldn't follow what they were saying." "It's insulting to be read to. I'm an adult and can read for myself." "I fell asleep." "I left before they finished." Reading a paper verbatim is universally a failure even though the presenter may have spent hours or weeks in preparation, might be very knowledgeable about the subject, and might even have beautiful slides. There are many reasons for this negative reaction.

First, it is unnatural and unusual for you to read to others. In your personal life and in your business life, you hardly ever write a script and then read it to your friends and co-workers. Furthermore, people rarely listen to reading, and so the audience has no feeling of rapport with the speaker. There is no eye contact, no feeling of human contact. Another problem with reading a speech is that we tend to write in a completely different manner than we talk. Most people write in nice, neat sentences and paragraphs, but we talk in short, brief bursts of spontaneous thought. Meanwhile, our ears have become accustomed to listening to talking, and our eyes have become accustomed to reading writing. If you have ever read a verbatim transcript of a trial or a meeting, you know that it looks like gibberish; when you put talking on paper, it is impossible to read. Conversely, when you try to listen to writing, it sounds odd. The ear is just not attuned to that style of communication. Therefore, your problem with the script is that when you write it out, it looks perfect to you, but it does not sound good to the audience. My conclusion is that you should make every effort to avoid writing a script.

Certainly there are exceptions. If the president of a company intends to make a speech to the stockholders or to the security analysts, and the primary purpose is to be quoted in the newspaper accurately, then certainly the president should write his or her speech. It would also be appropriate to have portions of your presentation scripted when they contain specific data that must be related in exact detail, but for most presentations, consider another method.

You will come across to your audience much better if you use the extemporaneous style, which means that you plan your ideas in advance, but use the words as they come to you conversationally. Using index cards with key words or a point-by-point outline of your material is one approach. However, if you do use the outline method, make sure that your printing is large and simple to follow.

A more professional and businesslike version of the extemporaneous presentation is one that includes visual aids, such as charts, slides, or an overhead projector. In what is sometimes called the "idea graph" method, you use the visual aids as your notes because they include the highlights of your presentation. You have the additional advantage of the audience being able to follow you readily because they can see the same highlights. I recommend the extemporaneous or the idea graph concept for nearly all of your presentations, including talks at conventions. You will sound much more natural to the audience, and they will be able to relate to you better and follow you more readily. You will find it more comfortable than reading a script, and you also will find it much easier to prepare. Writing a script is a very slow and tedious process. The following steps can be useful in preparing an idea graph type presentation.

Step 1. Develop an outline.

Step 2. Rough out your visual aids and key points on single sheets of paper (each visual aid should be on a separate sheet).

Step 3. Talk through these ideas with yourself or a fellow worker. A dry run will give you the opportunity to rearrange the sequence, adjust wording, etc.

Step 4. Create the final visual aids.

Some people, however, find it easier to reverse the sequence of Steps 1 and 2. Typically, your visual aids are the main points you want to make. You might find it easier to lay out all the main points first and then start connecting those with various words and emphasis points. Don't think about your opening until you get the entire presentation laid out. Too often, people spend an inordinate amount of time thinking about how they will open a presentation, and then regretfully realize they have very little time left to finish. Decide your main points first, then come up with an appropriate opening and an appropriate conclusion.

Logistics and Physical Arrangements

We all know Murphy's law, and it certainly is applicable on those occasions when you are expected to communicate to a group. Small, incidental details can ruin a presentation. On one occasion, the speakers at a gathering of about 150 people had been asked their requirements in advance. One speaker needed only a flip chart and a marking pen, two simple requests. On the day of the symposium, the sponsors had provided a flip chart, but they had given him a yellow marking pen, and it had not occurred to him to bring his own. Needless to say, writing with a yellow marking pen on white paper was disastrous.

On another occasion, a scientist was asked to make a presentation to the management committee of his company. This was a tremendous opportunity to convince them that his project was moving along well and that it was time for more funding and expansion. About an hour before the presentation, he was informed that the chairman of the board would also be sitting in, and the scientist knew that this would be one of the rare times he would be exposed to the top executive of the company. About two minutes before he was supposed to go on for his thirty-minute talk, someone whispered, "Bill, don't run late. The chairman must leave here at exactly 12 to catch a flight to New York." Bill replied, "No problem. I figure it will take me about thirty minutes." As he walked to the podium and turned around to face his audience, a sudden sinking feeling hit him, as he thought, "My entire career could depend on whether or not the lamp burns out in this projector." The point was that if the lamp did go out and people had to spend ten or fifteen minutes to find a spare, Bill's window of opportunity would vanish forever. That may be why in today's modern projectors, two lamps are now standard and they can be switched instantaneously. That is just an example of how little things can ruin a big presentation.

The purpose of these stories is to remind you how critical it is that you have all bases covered. Before every presentation you should go over a checklist such as the example shown.

Checklist for Physical Arrangements

Projector and spare lamp	Marking pen
Screen or clean white wall	Chairs and tables
Table for projector	Chalkboard and chalk
Microphone	People/invitation list
Conference room size	Podium
Samples/equipment	Refreshments
Handouts/literature	Time allotment
Agenda	Note pads, pencils
Name/place cards	Sign-in sheet
Extension cord/adapter	Chart pad and stand
Water and glass	Smoking and nonsmoking sections

This checklist may not be perfect for you, but you should have one, just like the pilot of a major airline. As he or she takes off for the 3000th flight, he or she will very carefully and very thoroughly examine a checklist.

You should also be concerned about possible competition during your presentation. Competition can come from outside noise, windows, telephones, discomfort, attention span, handouts, and time of day.

The first three items are probably obvious. Regarding discomfort, when people are uncomfortable, such as from sitting too long in one place or in very hard chairs, it is almost impossible for them to pay attention to you. Our feeling of self-preservation is instinctive, and our body comfort is closely related to it. If your audience is not comfortable, they will undoubtedly not pay much attention to you.

The average person's attention span is very short; people's minds tend to wander; and movement in the room is distracting. Be sensitive to the fact that just because you are mouthing the words does not necessarily mean that the audience is hanging on everything you say.

Handouts can be very serious competition. The next time you go to a meeting and the speaker gives a handout of several pages to the group, look around the room and watch what happens. When the speaker starts talking about the items on page one, the odds are that nearly everyone in the room is glancing through other pages. In fact every mind in the room may very well be on a different page or a different point. Unfortunately for the speaker, the human mind is not capable of concentrating on one area and listening to another.

Finally, the time of day can also be competition. For example, if you are part of an all-day program and you have been scheduled at 4:30 Friday afternoon, you are in trouble. If you are giving a one-hour slide presentation and you are scheduled at 1 o'clock, right after lunch, you are likely not to have an audience that's awake.

Whenever possible you should make any efforts you can to control all the environmental factors. You might want to consider these additional points:

Try to obtain a room of a size that will match the group size. One of the worst feelings is being a member of a small group sitting in an extremely large room. Most people get a feeling of being lost in space or a feeling of loneliness. Every time I give a presentation in that sort of atmosphere, there is no reaction and no laughter at the right places, just dead silence and probably inattention. For whatever reason, humans like to congregate close together. Perhaps you have noticed that the small bars and restaurants are always the most crowded and the most popular. On the other hand, a room that is much too small for its audience is also disastrous. Last minute and late arrivals scrambling for seating will disrupt any presentation. People trying to get comfortable or arrange their possessions will distract the speaker and each other.

Get as close as you can to your audience. If you are up on a stage behind a podium, removed from the audience, you cannot effect a feeling of rapport.

Make sure the background behind you is clean and neat. There should be no miscellany laying around and no writing on the blackboard from yesterday's meeting.

No other speakers should be up front with you unless it is a tandem presentation, because when you are making a key point somebody is likely to pour ice water or drop a book or create some such distraction.

Podiums have been a standard presentation tool for many years, but now they are probably used more out of tradition than for any real value. Podiums provide a convenient place to store things or to set notecards. They are even very good for leaning one's elbow on to relax. However, podiums create two problems: Some people get the feeling that they are being lectured or sermoned to because of the association with the pulpit. The other problem is that it is simply not natural for you to erect a barrier between yourself and another person when carrying on a conversation.

Think about the ways that you do converse with a few people in a more relaxed atmosphere. You may sit, you may move, you may talk with your hands, and so on. However, the script and the podium force you to be unlike yourself. This unnatural feeling is very disconcerting, particularly when you are under pressure. I prefer a tabletop podium that gives me a place to stand behind and so forth, but I also may walk around in front, occasionally sit on the edge of the table, perhaps even lean on the podium; the main point is that I don't get locked in behind it.

This podium problem is magnified in situations where you must use a microphone. The typical microphone is fixed to the podium, so that you cannot move your head more than about fifteen inches. How many times have you watched a speaker trying to point to a chart or screen and at the same time keep his or her mouth close to the microphone? Contortions like this are very distracting to the audience, disconcerting to the speaker, and not very effective. You should always use a lavaliere microphone, the kind that you can hang around your neck or clip on your collar, lapel, or tie. With a lavaliere you have total freedom and mobility; you can walk to a chart or a screen to point, and you can turn your head in any direction and not feel concerned about being close enough to the microphone.

Delivering the Presentation

Overcoming Stage Fright

Earlier I mentioned that speaking before a group is the most widespread fear of people in our country. Now I will discuss ways to overcome stage fright. A little stage fright is perfectly normal and healthy. It means that you want to do a good job, that you are concerned about being successful. Instead of worrying about eliminating stage fright, it is probably wiser to try to reduce it or at least hide it.

The first and most important rule is preparation. You must know your subject matter. If you have really prepared your material and are confident that you cannot be caught with an unanswerable question, you give yourself good reason to feel confident. You could also rehearse possible responses if you are caught off guard with a question you cannot answer.

The second rule is to dress your very best every time you have an important presentation to give. We all realize that other people tend to judge us by our appearance, and when you are dressed up you will naturally feel better.

Although I have mentioned the hazards of reading from a script, a good idea might be to have the first few sentences written out. Then, when you begin, you have those first important words and you don't have to worry about forgetting them. A good start will boost your confidence.

Many people believe that they should find a spot on the back wall and look at it during their presentation. It is not natural to stare at a thermostat or a wall switch for fifteen minutes. It is much more natural to look at a face, particularly a friendly face, and look at that person while you are addressing the entire audience. The feeling you will have is that you are talking to an individual, and after a few seconds of that, look at another face for a while and slowly move around the room from one face to another. Everyone is comfortable in one-on-one conversations.

Another technique to overcome nervousness is to do something physical at the beginning of your presentation. Pass out something before you start, or write a few words on a chart or a blackboard. Physical activity tends to reduce mental tension.

You might try a "dress rehearsal" with a friend, co-worker, or relative. The more you practice, the more confident you feel.

The Opening

A famous sage once said, "You never get a second chance to make the first impression", and that truism certainly applies to making presentations. Your opening is crucial to your success. One of the best techniques is complete silence. I have watched many speakers stand and wait and simply look at the audience while they were settling in their seats or finishing their whispering, and suddenly, a hush settles over the entire audience. If you've ever observed it, you know what I mean; it might work for you.

You could start by focusing temporarily on another person whose work might be appropriate to recognize. For example, "Before I begin, I would like to introduce my research assistants, Jean Snyder and Bob Jones", and perhaps add another sentence praising them or thanking them. Or it could be your boss whom you introduce. The purpose of this type of opening is to give you a chance to get comfortable mentally and physically.

Another idea for a good beginning is to start your presentation in a conversational style, such as by asking a question of the audience or an individual. Choose a question that you know they will answer in a positive manner.

The best-known opening for a presentation is telling a joke, a story, or a humorous incident. Many people succeed with that approach, but just because it succeeded for someone else does not necessarily mean it will work for you. Humor should be used only if you are certain that it will work, that your story is funny, and that everyone will laugh. If you are not a humorous person and you understand that about yourself, do not feel obligated to try to be funny. You can be very effective without using humor. On the other hand, if you do have an appropriate story that fits the occasion, nothing is better to break the ice and loosen everyone up, including yourself.

The humor must be appropriate. If you tell a racist, sexist, ethnic, or off-color joke, even if ninety percent of your audience laughs, you failed, because if you insult just one person, you failed. Your basic function in giving a presentation is to inform, not to entertain, and certainly not to insult.

One of the best techniques for getting the audience's attention is to have something that arouses their curiosity, such as an item covered with a cloth or a large box. The idea is to tantalize the audience into the question "What's he got there?" or "What's she up to?"

Many successful presenters open with a visual aid rather than words. Turning on a projector and letting the audience look at a chart or a graph without saying a word and letting the audience absorb that information sometimes says more than words can ever say.

The technique you choose will depend on the attitude of your audience toward you, your product, or your proposal. As I indicated earlier, there are three potential attitudes toward you: respectful, neutral, and defensive.

Obviously, with a respectful audience, you can get away with almost any kind of humor and not lose their respect. But, when you have a respectful audience, it is not humorous to be boastful about your talents or your performance; in fact, humility may be the best approach. For example, if Dr. Jerry Johnson is introduced as probably the world's finest biochemist, and then comes up to the platform and says, "Yes, it is true. I am the world's finest biochemist", he or she will probably create more bad will than good. On the other hand, if Dr. Johnson says, "I certainly appreciate those kind remarks, but the facts are that I have been very fortunate to be in the right place at the right time and I am sure that any of you could have accomplished what I have done if you had had the same opportunity", he or she will win over the audience immediately.

More often, your audience may be neutral or open-minded and waiting to see and willing to listen to what you have to say. Humor is very appropriate in this situation, but you have to be certain that it will go over well, and you must avoid insulting types of jokes. With a neutral audience, the humble approach is once again more appropriate than the boastful approach, but you should give a brief background on your work or your experience, and also give the audience justification for listening to you. Why should they bother to pay any attention unless there is some benefit in it for them? Be sure to inform them of such a benefit in your opening.

If your audience is defensive, you might as well not even try any humor, because every time you do, they will be thinking, "He's trying to warm me up. Well, he's not going to." Only one thing works with a defensive audience: immediately lay all the cards on the table. Step 1, you admit your differences. Step 2, you present their favorable point of view. Step 3, appeal to their objectivity. A simple example might be as follows. "I am well aware that most of the people present today are opposed to spending any more money on this project. Frankly, if I were in your shoes, I would probably feel the same way, because it does seem to be foolish to continue to pour more money into a project that has been nothing but failure so far, and there are lots of other areas where the money could be spent wisely. All I ask is that you give me fifteen minutes to explain some new developments that have taken place that might shed a new light on your perspective. I ask you to weigh these developments before you make your final decision."

Making Your Message Clear

To reiterate, communication is not necessarily what we say to people, but rather, their interpretation of our words. When you give a presentation, you have to make sure that the message you want to convey is totally clear to your audience. The responsibility and the burden of accurate communication are on you, not them. One of my favorite sayings is "The biggest barrier to communication is the assumption that it took place." We say things and assume that people know what we mean. When they hear us, they assume that they know what we mean. And yet all too often, we find ourselves disagreeing about what was said only a few hours or a few days earlier. Conflict, misunderstandings, and confusion almost always result from poor communication.

To make yourself clearly understood, you may have to resort to techniques such as stories or parables, and give extra details that allow the listener to create a mental picture. Facts by themselves are sometimes not clear. If you state a fact that you know specifically to be true, it may have no meaning to your audience. As an example, if you were trying to explain to an audience that the state of Alaska was very large, and you said that it contained 497,000 square miles, you would be stating the truth. You would also be stating a fact that has no meaning because the average person has no idea how large a square mile is or how large his or her own state is. On the other hand, if you said, "The state of Alaska is larger than Texas and California combined", your meaning is much more clear. That word picture makes sense to the average person; it is a comparison that most people understand.

Finally, be very sensitive to your own special vocabulary and the jargon in your field. Be careful not to use terms that your audience may not understand.

Answering Questions

From the audience's point of view, how you answer questions may be a critical part of their measurement and evaluation of your presentation. You may not think the question is particularly important, but they certainly do, or they would not ask. Therefore, keep in mind the following rules when you answer a question from an audience.

Listen to the question. That means eye contact and slightly forward body language that suggests total attention and interest. If you start searching through a briefcase or a file folder while your questioners are talking, the audience has the feeling that you are not hearing them.

Acknowledge or compliment the question. Statements like, "That's an excellent point" or "I'm glad you brought that up" are complimenting and reassuring to the person who asked.

Repeat the question, especially if the question comes from the front of the room. As you repeat the question, be sure that you state the intent of the question, not its literal wording. Sometimes, when people have less knowledge of your subject than you do, they may have difficulty with phrasing; your job is to help them. After you have answered the question, you should ensure their satisfaction with queries like, "Does that answer your question?" or "Does that clear it up?" If people are not satisfied that you have answered their questions, that dissatisfaction will be on their minds and they will tend to tune you out.

The Conclusion

Any good presentation has a conclusion, not just an end to the speaking. The key points to cover are the three Rs: recap, restate benefits, recommend. In other words, summarize the highlights; restate the benefits derived from this presentation, such as time or money saved; and make your recommendation.

Visual Aids

Your odds of being successful in any presentation are greatly enhanced if you support yourself with some type of visual aid. First, consider what a visual aid tells the audience about you. Visual aids indicate that you are well prepared, that you are serious, and that you know what you are talking about. If you have taken the time and effort to make visual aids, you certainly have studied your subject; therefore, the visual aids you use must be accurate. In addition, visual aids should do the following for you: (1) help you gain and hold attention, (2) keep you organized, (3) give you greater confidence, and (4) allow you to maintain better control.

With visual aids, you will communicate more accurately and clearly because people understand more and retain more when they see as well as hear the main parts of a message. Recent studies indicate that we retain only about 20% of what we hear, 30% of what we see, 50% of what we see and hear, and 90% of what we see, hear, and do. Therefore, using visual aids in a presentation in which there is a great deal of participation is by far the most effective approach.

Most audiences are made up of people who have relatively short attention spans, are not particularly good listeners, and tend to think much faster than you can talk. Therefore, their minds wander. By using visual aids, you can focus attention on the point you want to make and have a much greater likelihood of holding attention.

Which type of visual aid is best for you? The following is a quick look at the advantages, disadvantages, and best possible uses for the four major visual tools.

Chalkboards are spontaneous, easy to use, inexpensive, and erasable. However, writing on a chalkboard is slow, often not readable, and temporary. Furthermore, you have to turn your back to the audience while you are writing, and, of course, you must have chalk. Chalkboards are best used at small, informal meetings, brainstorming sessions, or meetings where you are trying to generate spontaneous ideas as you speak.

Prepared charts should have better contrast than a chalkboard, and they are permanent, so you can refer to them many times. Because you prepare them in advance, you need not keep the audience waiting while you write, as with a chalkboard. However, they are bulky because they must be large enough for the entire audience to read. Also, they can be expensive if you have them prepared professionally. Their best use is with small groups and for brainstorming sessions.

The highest quality visual aid is the 35-millimeter slide. Only with slides can you show color photographs that have a lot of detail. Slides are very portable, and a slide projector is easy to operate and appropriate for any size group. However, you must be able to darken the room. Slides are usually expensive and can become the focal point of the presentation instead of the speaker. Slides are best used in repetitive programs where photographic detail and a professional look are important. Also, slides are excellent for training programs or product displays, and sound or music can be added.

In my opinion, nothing can quite compare with the overhead projector. It is the only one that is totally flexible and allows you to be completely natural when giving a presentation. It is appropriate for any size group. Furthermore, when you use an overhead projector, you face your audience. When people come to your home or your office, you don't turn your back on them, do you? You can use a pen or a marker as a pointer, or, in addition to the artwork you are displaying, you can write on the stage of the projector. It is very natural to sketch and point as we talk. Thus, you can combine artwork prepared in advance with spontaneous ideas.

A transparency should be treated just like a sheet of paper that you are sharing with a much larger group. Transparencies themselves can be made quickly and easily from any document that you have at your disposal because of the proliferation of plain-paper copiers.

This ease of preparation of transparencies leads to their main disadvantage: Transparencies are so simple to make that people get lazy and copy just anything at hand, which often is a typewritten sheet, a computer printout, or a very fine line drawing. None of these really qualifies as a good visual aid. Furthermore, the projector can block some people's view if it is not positioned carefully, and it is much less portable than a 35-millimeter projector.

The overhead projector is best used in company meetings, for financial and technical presentations, for group sales presentations, and at seminars or workshops where the speaker wants to maintain rapport.

Whatever you choose, make sure that your visual aids are large and readable by your entire audience. Also, visual aids should be brief, simple, and to the point; they should contain key words instead of sentences. Furthermore, if you have not had much experience with the device you will be using, practice using it and become comfortable with it before you give your presentation.

Conclusion

If nothing else, I hope that I have conveyed to you the feeling that presentations are a key to your growth and that each one, regardless of audience size, is an opportunity to have others take notice of you. Therefore, it is important that you take the time to put extra effort into your planning and into your presentation.

Appendix I

ACS Publications

Editorial Procedures

In ACS books and journals, the specifics may differ from office to office, but the general procedures for processing manuscripts from review through production are similar. More importantly, the philosophy is the same.

Manuscript Review

Papers submitted to ACS books and journals are considered for publication with the understanding that they have not been published or accepted for publication elsewhere and are not currently under consideration elsewhere. Appendix II, "Ethical Guidelines to the Publication of Scientific Research", presents ACS's guidelines for editors, authors, and reviewers.

The following consensus statement of ACS editors represents current peer-review practices of ACS journals and books:

> Manuscripts submitted for publication in ACS research journals and in ACS book series, after screening by editors for scope appropriate to the intended publication, are generally reviewed by two or more individuals. Reviewers are chosen on the basis of their known expertise in the research field covered by the manuscript. The ability of an individual to render an objective, critical review is also considered in reviewer selection.

> Among the criteria reviewers are asked to use in judging manuscripts are originality of the work, its significance to the research field, the validity of experimental data, the adequacy of supporting information, and the soundness and logic of interpretation. The specific criteria used vary according to the particular editorial requirements of each journal or book.

> Authors are sent pertinent reviewer comments by the Editor [journal editor or ACS Books staff editor in conjunction with the book editor], generally with the identity of the reviewer removed. The final decision to publish is the responsibility of the Editor. Although decisions are nearly always based on reviewers' comments, each editor retains the right to publish in the face of negative comments and to reject in the face of positive comments. In practice, however, editorial decisions that are contrary to reviewer recommendations are usually related to journal or book scope or involve manuscripts in the Editor's particular field of expertise.

In ACS books, rejections may be made on the basis of timing and scheduling if a manuscript will require extensive revision and additional review. Further review often is not possible because of time constraints. In journals, a paper that is rejected may be reconsidered if the author presents a good case for further review.

A paper may be rejected if the author refuses to obtain the necessary permissions to reproduce previously published material (such as figures or tables) or if the author refuses to sign the ACS Copyright Status Form.

The actual sending of the manuscript for review and receiving of reviews is done by staff in the individual journal offices or the ACS Manuscript Office, depending on the journal. The ACS Books Department handles reviews of manuscripts submitted for most books. ACS Symposium Series manuscripts are the exception; reviews are handled by the book editor.

After a revised paper is accepted, it is considered to be in final form. Alterations made after acceptance may be permitted at the discretion of the editor. If alterations are extensive or if significant additions are made, additional review may be required. This may not be possible in ACS books.

Processing of Accepted Manuscripts

Manuscripts accepted for journals go directly from the editor or managing editor to the Journals Editorial Department, where they are copy edited and prepared for the printer. Manuscripts accepted for *CHEMTECH* and the A pages of *Analytical Chemistry* and *Environmental Science & Technology* are copy edited and prepared for printing by staff in the ACS Graphics and Production Department. All manuscripts accepted for books are prepared for the printer in the Books Department.

Technical Editing Technical editing of papers is quite extensive, and involves both production and copy editing. Routine production consists of scheduling, marking the copy according to publication style, selecting or specifying type, preparing page layout, designing book jackets, sizing art, reviewing or proofreading galleys, checking page makeup, giving instructions to typesetters and printers, and preparing indexes. Copy editing is done to ensure consistency, clarity, and grammatical accuracy; changes are introduced to improve nomenclature, graphics presentation, and tabular format. Copy editors may contact authors or write queries on the galleys to request clarification of material.

Author's Proof One author, generally the author to whom correspondence should be addressed (or the first author listed if not otherwise specified), receives a galley proof of the manuscript for final approval before publication. A manuscript usually is not released for printing until the author's proof or other approval (e.g., via telephone) has been received by the Journals

Editorial, Graphics and Production, or Books Department. Hence, galleys should be checked and returned promptly according to individual journal or book instructions. To save time and expense, foreign contributors may authorize a colleague in the United States to read galleys.

Authors should check galleys thoroughly by reading them at least twice. Only corrections and necessary changes can be made in galleys. Extensive changes will require editorial approval and a revised date of receipt and will delay publication. Printer's errors are corrected at no cost to authors, but authors may be charged the cost of extensive resetting made necessary by their own alterations. In books, there simply may not be time to do extensive alterations at all.

Instructions to Authors for Proofing Galleys   Other points the author should keep in mind when proofreading are as follows:

- Mark corrections in the margins of the galleys in pencil.

- Do not erase or obliterate type; instead, strike one line through the copy to be deleted and write the change in the margin.

- Clarify complicated corrections by rewriting the entire phrase or sentence in the margin.

- Check all text, data, and references against the original manuscript. Pay particular attention to equations; formulas; tables; captions; spelling of proper names; and numbering of equations, illustrations, tables, and references.

- Answer explicitly all queries made by the technical editor.

Corrections to Published Manuscripts

Corrections for a paper that has already been published should be sent in duplicate to the editor or managing editor of the journal or the Books Department. Most ACS journals publish corrections soon after they have been received. The Books Department will print an Erratum sheet to be included in every book, and the book itself will be corrected before reprinting.

If authors notify the Editorial or Books Department promptly, reprints may be corrected. When such corrections are possible, the corrected reprint will serve as the master from which the microfilm version of the journal is prepared.

Reprints

Reprints should be ordered on the appropriate form, which is sent to authors with galleys of journals or shortly after galleys of books. Authors should follow the instructions on the forms themselves. Usually, a purchase order should be submitted with the order.

Liability and Copying Rights

Authors are solely responsible for the accuracy of their contributions. The ACS and the editors assume no responsibility for the statements and opinions advanced by the contributors.

Contributions that have appeared or have been accepted for publication with essentially the same content in another journal or book or in some freely available printed work (e.g., government publications, proceedings) will not be published in ACS publications. This restriction does not apply to results previously published as communications to the editor in the same or other journals.

Books

Advances in Chemistry Series

Books in this series are collections of papers on a particular topic. They may be developed from symposia or they may be planned specifically for book publication from the outset. Papers are reviewed critically, and the same rigorous standards of acceptance that apply to papers for ACS journals are applied to papers for these books. Papers may be reviews or reports of original research with review material included.

ACS Symposium Series

This series was originated to provide more timely publication of original research papers presented at symposia. Books in this series usually cover rapidly developing areas of chemistry, for which time of publication is especially important. Each paper undergoes limited peer review. This review is conducted by the editor of the book with specific guidelines provided by the Books Department. After the papers have been reviewed and revised, the authors prepare camera-ready copy, from which the book is produced.

ACS Monographs

Monographs are intended to serve two principal purposes: (1) to provide chemists with a thorough treatment of a selected area in a form usable by experts as well as persons working in tangential fields and (2) to stimulate further research in the specific field treated.

Other Books

ACS also publishes books that are not part of any particular series. These include reference, data, and reprint volumes as well as books that are semitechnical or nontechnical in nature. Often these books appeal to a broader audience than the technical books in the three series. Most of these books undergo stringent peer review.

ACS publishes three directories: (1) *College Chemistry Faculties*, which lists faculty members in schools granting associate, bachelor's, master's, and doctoral degrees in chemistry in the United States and Canada; (2) the *ACS Directory of Graduate Research* (DGR), which lists faculty members, their areas of expertise, and their recent publications for schools granting master's and doctoral degrees in the United States and Canada; and (3) *Chemical Research Faculties: An International Directory*, which is the international complement to the DGR and covers the remainder of the world. Currently the DGR is also available online, and plans are underway for making *Chemical Research Faculties* available online also.

Journals

Accounts of Chemical Research

This journal publishes concise, critical reviews of research areas currently under active investigation. Most articles are written by scientists personally contributing to the area reviewed. Reviews need not be comprehensive. Indeed, they may be concerned in large part with work in the author's own laboratory. Most reviews are written in response to invitations issued by the editor. First- and third-person nominations of prospective authors are welcomed; nominations should suggest the research area for review, and they should briefly summarize the prospective author's contributions to it. Unsolicited manuscripts are also considered for publication.

Analytical Chemistry

This research journal is devoted to all branches of analytical chemistry. Research papers are either theoretical with regard to analysis or are reports of laboratory experiments that support, argue, refute, or extend established theory. Research papers may contribute to any of the phases of analytical operations, such as sampling, preliminary chemical reactions, separations, instrumentation, measurements, and data processing. They need not necessarily refer to existing or even potential analytical methods in themselves but may be confined to the principles and methodology underlying such methods. Critical reviews of the literature, prepared by invitation, are published in April of each year in special issues that cover, in alternating years, applied and fundamental aspects of analysis.

Biochemistry

This journal publishes the results of original research that contribute significantly to biochemical knowledge. Preference is accorded to manuscripts that generate new concepts or experimental approaches, particularly in the advancing areas of biochemical science. Hence, the primary criterion in the acceptance of manuscripts is that they present new and germinal findings.

Chemical & Engineering News

As the official publication of the American Chemical Society and as a chemical news weekly, C&EN is designed to keep ACS members informed of policies and activities of the ACS, to keep members as well as other readers well informed on the activities of the chemical world through news reporting and through calling attention to issues of consequence to chemists and chemical engineers, and to relate the beneficial contributions of ACS and its members to the broad goals of society at large.

Chemical Reviews

Articles published in this journal are authoritative, critical, and comprehensive reviews of research in the various fields of chemistry. Preference is given to creative, timely reviews that will stimulate further research with a carefully selected subject and well-defined scope. In general, the topic should not have been reviewed in a readily available publication for about five years, although exceptions will be made if developments in the field have been particularly rapid or if new insight can be achieved through further review of the subject. Reviews are invited by the editor in response to suggestions from the editorial advisory board or elsewhere in the scientific community. Articles may also be accepted from authors who wish to contribute an unsolicited paper, provided that they contact the editor and obtain preliminary approval of the project, according to the procedure outlined in the "Suggestions to Authors" published in the journal.

CHEMTECH

Every month CHEMTECH, "The Innovator's Magazine", makes accessible a variety of useful tools for the mission-oriented chemist and engineer. Its concept orientation treats not only preparation, characterization, and uses of materials but also process design, data reduction, and, most significantly, all the sister areas important to bring products to society: economics, law and regulation, marketing, management in general, and, of course, education. Above all there is focus on concerns of the responsible technologist. Each article is written by a recognized authority. Case history and workbook styles and multidisciplinary approach stimulate the broad-base awareness and interest that creative solutions to real problems require. Each issue treats the chemical technology of our vital needs—materials, energy, food, clothing, shelter, transport, medicine, and the like—as well as the interaction of individual sciences and technologies with each other and with socioeconomic issues of the present and future.

Environmental Science & Technology

This journal/magazine places special emphasis on reporting original chemical research, engineering developments, and technicoeconomic studies in fields of science directly related to the environment. Contributed articles are directed to scientists and engineers concerned with fundamental and applied aspects of water, air, and waste chemistry. Because a meaningful approach to the management of environmental quality involves more than scientific understanding, the journal devotes serious attention to engineering, economic, legal, and other influences to give its readers an integrated view of this complex system. Contributed papers should describe results of original research. Review articles are considered when they serve to provide new research approaches or stimulate further worthwhile research in a significant area. The journal also publishes correspondence and contributed scientific features on topics of current interest.

Industrial & Engineering Chemistry Fundamentals

Papers in the broad field of chemical engineering research are presented here. No technical field to which important contributions are being made is excluded—whether experimental or theoretical, mathematical or descriptive, chemical or physical. Acceptable papers are characterized by conclusions of some general significance, as distinguished from papers intended mainly to record data. Treatment of subjects is based on application of chemical, physical, and mathematical sciences, and papers are judged on their perceived lasting value. Purely mathematical articles are acceptable only when they contain significant conclusions that relate to real chemical engineering problems.

Industrial & Engineering Chemistry Process Design and Development

Reports of original work on design methods and concepts and their application to the development of processes and process equipment are published in this journal. Empirical or semitheoretical correlations of data, experimental determinations of design parameters, methods of integrating systems analysis and process control into process design and development, scale-up procedures, and other experimental process development techniques are included.

Industrial & Engineering Chemistry Product Research and Development

Papers are published under these headings: Reviews, Symposia Papers, Catalyst Papers, and General Papers. The reviews are in-depth analyses covering product areas of major interest. Symposia topics include blocks of selected papers from product-related symposia worldwide. Catalyst papers broadly encompass catalysis, with additional emphasis on automobile exhaust catalysis. The function of general papers is to describe the results of recent research in fields as diverse as coatings, plastics, agriculture, communications, papermaking, textiles, fuels, and polymers.

Inorganic Chemistry

This journal publishes original studies, both experimental and theoretical, in all phases of inorganic chemistry. These include the synthesis and properties of new compounds, quantitative studies reporting thermodynamic and kinetic studies related to reaction mechanisms, structural studies, and all aspects of bioinorganic chemistry. Short papers (Notes), full papers, and preliminary communications of an urgent nature are published.

Journal of Agricultural and Food Chemistry

This journal publishes research findings from the several interrelated chemical fields closely associated with the production, processing, and utilization of foods, feeds, fibers, and forestry products. Topics most frequently discussed are pesticides; fertilizers; plant growth regulators; the chemical composition of flavors, food materials, and other plant materials; the chemistry of nutrition; and the processing of food and farm needs. The papers submitted should report original research, and some practical significance should be apparent.

Journal of the American Chemical Society

Original papers in all fields of chemistry are published here. Emphasis is placed on fundamental chemistry. Papers of general interest are sought, either because of their appeal to readers in more than one specialty or because they disclose findings of sufficient significance to command the interest of specialists in other fields. Communications are restricted to reports (usually preliminary) of unusual urgency, significance, and interest, and are limited to 1000 words or the equivalent. Book reviews are also published. Specialized papers are not published.

Journal of Chemical and Engineering Data

This journal is directed to the publication of experimental or, in some cases, derived data in sufficient detail to form a working basis for applying the information to scientific or engineering objectives. Experimental methods should be referenced or described in enough detail to permit duplication of the data by others. The data should be presented with such precision that the results may be readily obtained within the stated limits of uncertainty of the experimental background. For most studies a tabular presentation or a mathematical description is preferred to the use of graphical methods.

Journal of Chemical Information and Computer Sciences

This quarterly journal publishes research papers in all areas of information and computer science relevant to chemistry and chemical technology. Feature articles and book reviews help the reader stay current in this rapidly changing technological area. Subject fields include data acquisition and analysis, pattern recognition, artificial intelligence, new algorithms and applications of existing algorithms, interfacing techniques, computer systems and components, graphics, simulation and modeling, operations research, econometric and management systems, indexing classification and notation systems, data correlation information systems, nomenclature, linguistics and communication, information sources and services, and design and applications of computerized information and data systems.

Journal of Medicinal Chemistry

This journal publishes articles, notes, and communications that contribute to an understanding of the relationship between molecular structure and biological activity. Some of the specific areas that are appropriate are analysis of structure–activity relationships by a variety of approaches; design and synthesis of novel drugs; improvement of existing drugs by molecular modification; biochemical and pharmacological studies of receptor or enzyme mechanisms; isolation, structure elucidation, and synthesis of naturally occurring, biologically active molecules; new, improved syntheses of important drugs; physicochemical studies on established drugs that may furnish some insight into their mechanism of action; and the effect of molecular structure on the metabolism, distribution, and pharmacokinetics of drugs. In addition, interpretive accounts (PERSPECTIVE Articles) of subjects of active current interest are published at the invitation of the editors.

The Journal of Organic Chemistry

The aim of this journal is to publish original and significant contributions in all branches of the theory and practice of organic chemistry. Areas emphasized include the many facets of organic reactions, natural products, bioorganic chemistry, studies of mechanism, theoretical organic chemistry, organometallic chemistry, and the various aspects of spectroscopy related to organic chemistry. This journal publishes articles, notes, and communications.

Journal of Physical and Chemical Reference Data

This journal is published quarterly by the American Chemical Society and the American Institute of Physics for the National Bureau of Standards. The objective of the journal is to provide critically evaluated physical and chemical property data, fully documented as to the original sources and the criteria used for evaluation. Critical reviews of measurement techniques, whose aim is to assess the accuracy of available data in a given technical area, are also included. The journal is not intended as a publication outlet for original experimental measurements such as are normally reported in the primary research literature, nor for review articles of a descriptive or primarily theoretical nature.

The Journal of Physical Chemistry

This biweekly journal reports new experimental and theoretical research dealing with fundamental aspects of physical chemistry and chemical physics. Each issue begins with the Letters (short article) section where new findings that deserve rapid dissemination are reported. Invited Feature Articles by accomplished scientists working in fields of current interest to physical chemists and chemical physicists regularly follow the Letters. In Feature Articles, authors review their fields, with emphasis on their own contributions, for physical chemists who are not necessarily in the author's research area. A broad selection of Articles and occasional Comments complete each regular issue. Proceedings of selected symposia are sometimes published.

Langmuir

Langmuir is a new ACS bimonthly journal covering the broad area of surface-colloid chemistry. Topics range as follows: ultra-high-vacuum studies of the solid interface and of chemistry and electrochemistry at well-defined solid surfaces, heterogeneous catalysis, the liquid–vapor interface including capillarity, nonmolecular films, wetting and contact angle, and the large domain of disperse systems. Original research articles and Letters are accepted. There are occasional invited review-type articles and book reviews. The journal is intended to provide a unifying influence on a subject otherwise covered by many specialized and usually private journals.

Macromolecules

This journal publishes original research on all fundamental aspects of polymer chemistry, including synthesis, polymerization mechanisms and kinetics, chemical reactions, solution characteristics, and bulk properties of organic polymers, inorganic polymers, and biopolymers. The editors welcome regular articles, notes, communications, and occasional reviews.

Organometallics

Organometallics publishes articles and communications concerned with all aspects of organometallic chemistry. Specifically, it covers (1) synthesis; (2) structure and bonding—experimental and theoretical studies; (3) chemical reactivity and reaction mechanisms of the metal–carbon bond as well as of other inorganic and organic functionality present in the molecule; and (4) applications—organometallic reagents in organic synthesis and in polymer synthesis, catalytic processes in which an organometallic compound is the catalyst or in which organometallic species are intermediates, and organometallic compounds in the synthetic aspects of materials science and solid-state chemistry.

Microform Supplements

Manuscripts often contain material that, although essential to the specialized reader, need not be elaborated in the paper itself. Examples of such material are extensive tables; graphs; spectra; mathematical derivations; computer algorithms; protein sequence analyses; multiple regression analyses; experimental material not central to, but bordering on, the central theme of the paper; and expanded discussions of peripheral points. In journals, this material is made part of the permanent record by publication in the microfilm edition of the journal. In addition, the material is also provided on microfiche to subscribers of supplementary material (many journals offer these subscriptions) or on microfiche and photocopy by special order to any interested reader. The supplementary material is indexed by *Chemical Abstracts*.

Microform material should be submitted, in the appropriate number of copies and clearly labeled, with the original manuscript. After the paper is accepted, this material will be processed as is (unless the preparation requirements are not met), and proof will not be sent to authors.

Authors who use other documentation services or depositories (such as Cambridge Crystallographic Files) are encouraged to ask that this material, when relevant to the manuscript being considered for publication, be accepted as supplementary material. It will then become part of the permanent archives of the ACS by being included immediately following the article itself in the microfilm issue of the journal.

ACS Single Article Announcement

This is a semimonthly current-awareness service based on twenty ACS journals and magazines. The announcement consists of the tables of contents from the latest issues of these journals, and a single copy of any item listed may be ordered.

ACS JOURNALS ONLINE

The ACS JOURNALS ONLINE data base contains the complete texts (not tables or figures) of the nineteen primary ACS journals. Any part of a journal article can be searched and displayed; every word is searchable. Of particular interest is the capability to search for very specific kinds of information such as spectroscopic techniques, molecular formulas, thermodynamic data, toxicity data, biological data, and chemical names. More than 50,000 complete journal articles are included, dating back to 1980, and the data base is updated every two weeks.

Audio Courses

ACS Audio Courses are study units at the college and continuing education levels in chemistry, chemical engineering, and related subjects. Prepared by leading authorities and carefully edited by ACS staff, the courses consist of two components: a lecture portion on audiocassettes and an integrated reference manual that, when annotated by the listener, becomes a personalized state-of-the-art treatise. The catalog of courses includes both review and cutting-edge topics in science and technology, as well as nontechnical personal and professional development subjects that can play an important role in career advancement.

Chemical Abstracts Service

Chemical Abstracts Service (CAS), in its mission of providing access to the world's literature related to chemistry and chemical engineering, offers a wide range of publications and services. These include the comprehensive printed *Chemical Abstracts* (CA); an online chemical data base, CAS ONLINE; and a broad variety of other printed, microform, and electronic services intended to satisfy current-awareness and retrospective searching needs. The CAS data base is made up of abstracts of documents in the scientific literature, bibliographic citations, comprehensive substance-related information, and extensive index entries. It draws upon articles that appear in 12,000 scientific and technical journals published in 140 nations; patents issued by 28 patent offices around the world; and material from conference proceedings, technical reports, dissertations, and new books. All abstracts are in English, although the original literature may be in any of 50 languages.

Chemical Abstracts

Each issue of CA contains abstracts, bibliographic citations, and indexes. The abstracts are grouped into 80 subject sections. CA Sections 1-34 are published one week, and Sections 35-80, the next. The indexes in the weekly issues include Keyword Index, Author Index, and Patent Indexes. The 52 weekly issues of CA are divided into two semiannual volumes. Six volume indexes (General Subject, Chemical Substance, Formula, Author, Patent, and Index of Ring Systems) are produced. The Index Guide is a cross-reference tool that links substance and subject terms in general use with the highly controlled CAS terminology. Collective indexes combine into a single listing the content of the individual CA volume indexes for a five- or ten-year collective indexing period.

CA Section Groupings

CA Section Groupings divide the content of the weekly CA issues into five separate printed publications in related subject areas: Biochemistry; Organic Chemistry; Macromolecular Chemistry; Applied Chemistry and Chemical Engineering; and Physical, Inorganic, and Analytical Chemistry. Each contains abstracts and bibliographic information reproduced exactly from specific CA Sections as well as a keyword index. Each Section Grouping is published once every two weeks and corresponds to the issue of CA from which it is taken.

CA SELECTS

CA SELECTS is a series of about 150 current-awareness bulletins published every two weeks, 26 times per year. Each CA SELECTS topic is a separate publication. Each topic includes CA abstracts and associated bibliographic information, selected by computer from the CA data base according to a precise, special interest profile.

CAS BioTech Updates

CAS BioTech Updates includes abstracts from *Chemical Abstracts* and business information from *Chemical Industry Notes* related to biotechnology processes, people in the industry, government activities, production, and pricing internationally. Each issue is divided into four sections: patents, papers, books and reviews, and BioTech Industry Notes. Each issue also contains author indexes.

BIOSIS/CAS SELECTS

This series of biweekly current-awareness publications drawn from the BioSciences Information Service and Chemical Abstracts Service data bases covers the literature of biological chemistry, pharmacology, biochemistry, medical chemistry, and other fields related to chemistry and biology. Each topic provides pertinent abstracts or content summaries and bibliographic citations from the scientific and technical literature covered by *Biological Abstracts*, *Biological Abstracts/RRM* (Reports, Reviews, Meetings), and *Chemical Abstracts*.

Chemical Titles

This alerting service is published every two weeks and provides titles of articles appearing in 800 leading scientific journals published worldwide. Each issue contains a keyword-in-context index (arranged in alphabetical order), bibliography (arranged alphabetically by journal CODEN with abbreviated journal titles provided), and author index (arranged in alphabetical order).

Chemical Industry Notes

Chemical Industry Notes is a current-awareness service covering activities in the chemical industry and related industrial fields. Published weekly, this publication monitors nearly 80 leading business and trade journals worldwide. Each issue covers such areas as industrial management, investment, marketing, pricing, and production with brief but informative article extracts, bibliographic citations, and keyword and corporate indexes.

CASSI Cumulative 1907–1984

Chemical Abstracts Service Source Index (CASSI) *1907–1984* provides complete, information on serials and nonserials that are monitored by CAS and held by more than 350 libraries around the world (300 in the United States). It contains complete bibliographic data for nearly 60,000 publications, including information about variant titles, histories of publications, English translations of many foreign-language titles, a directory of publishers and sales agencies, and guides to the depositories of unpublished works. In addition, CASSI indicates which titles are available from the CAS Document Delivery Service. CASSI is updated with quarterly supplements. A microform CASSI KWOC (keyword-out-of-context) Index is available to determine full titles for publications by use of any significant word from the title.

Ring Systems Handbook

The *Ring Systems Handbook* (RSH) allows access to nearly 60,000 ring and cage systems formerly contained in the *Patterson Ring Index* and the *CAS Parent Compound Handbook*. The RSH consists of four volumes: a two-volume *Ring Systems File*, an index, and a cumulative supplement. In the *Ring Systems File*, entries are arranged in ring analysis order. For ring systems having a common ring analysis, ring names are listed in alphabetical order. Each entry is identified by a unique Ring File number. Structure diagrams for cage systems are grouped in ascending molecular formula order at the end of the *Ring Systems File*. The index provides two routes of access to ring information: the Name Index, where names are listed alphabetically along with a Ring File number, and the Ring Formula Index, where formulas are listed in Hill System order and include the Ring File number.

Registry Handbook—Common Names

This microform listing links a common name for a substance with a CAS Registry Number, as well as CA index name, molecular formula, and related names. Coverage includes over 1,100,000 names and 473,000 Registry Numbers.

Registry Handbook—Number Section

The Registry Handbook—Number Section provides the CA Index names and the molecular formulas for more than 7 million substances. The "base book" covers 1965-1971. Additions are provided in annual supplements based on specific Registry Number ranges.

International CODEN Directory

This microfiche index lists CODEN (six-character codes for publication titles) and full publication titles for 160,000 serial and nonserial publications. Access to these publications is available through CODEN title, or keyword-out-of-context indexes.

CAS ONLINE

CAS ONLINE is a comprehensive chemical information data base, available through the facilities of STN International, offering substance-oriented and subject-oriented searching. The data base includes three related files: the Registry File for substance identification, the CA File for bibliographic searching, and the CAOLD File for references to pre-1967 literature.

The Registry File contains information on more than 7 million substances reported in the literature; 10,000 new substances are added every week. The Registry File is the world's largest file of substance information, including coordination compounds, polymers, alloys, mixtures, and minerals. The Registry File may be searched by molecular structure; substructure; or a variety of chemical dictionary terms such as chemical names, name fragments, or molecular formulas.

The CA File contains bibliographic data and CA index entries for more than 6 million documents cited in CA since 1967. In addition, the CA File contains complete CA abstract text for documents abstracted since mid-1975 (and for some earlier documents). Both bibliographic references and index information can be searched and displayed in the CA File; by the end of 1985, CAS ONLINE will also have searchable abstract text.

The CAOLD File contains references to substances cited in CA before 1967. (Many of these substances also have references after 1967.) This file is being expanded incrementally as these "earlier" substances are added to the Chemical Registry. Only a limited amount of information is available for these pre-1967 references: the CA Accession Number, Document Type (for patents), and Registry Numbers as index terms.

Appendix II

Ethical Guidelines to Publication of Chemical Research

Preface

The American Chemical Society serves the chemistry profession and society at large in many ways, among them by publishing journals that present the results of scientific and engineering research. Every editor of a Society journal has the responsibility to establish and maintain guidelines for selecting and accepting papers submitted to that journal. In the main, these guidelines derive from the Society's definition of the scope of the journal and from the editor's perception of standards of quality for scientific work and its presentation.

An essential feature of a profession is the acceptance by its members of a code that outlines desirable behavior and specifies obligations of members to each other and to the public. Such a code derives from a desire to maximize perceived benefits to society and to the profession as a whole and to limit actions that might serve the narrow self-interests of individuals. The advancement of science requires the sharing of knowledge between individuals, even though doing so may sometimes entail foregoing some immediate personal advantage.

With these thoughts in mind, the editors of journals published by the American Chemical Society now present a set of ethical guidelines for persons engaged in the publication of chemical research, specifically, for editors, authors, and manuscript reviewers. These guidelines are offered not in the sense that there is any immediate crisis in ethical behavior, but rather from a conviction that the observance of high ethical standards is so vital to the whole scientific enterprise that a definition of those standards should be brought to the attention of all concerned.

0917-0/86/0217$06.00/0   © 1986 American Chemical Society

We believe that most of the guidelines now offered are already understood and subscribed to by the majority of experienced research chemists. They may, however, be of substantial help to those who are relatively new to research. Even well-established scientists may appreciate an opportunity to review matters so significant to the practice of science.

Formulation of these guidelines has made us think deeply about these matters. We intend to abide by these guidelines, strictly, in our own work as editors, authors, and manuscript reviewers.

Guidelines

A. Ethical Obligations of Editors of Scientific Journals

1. An editor should give unbiased consideration to all manuscripts offered for publication, judging each on its merits without regard to race, religion, nationality, sex, seniority, or institutional affiliation of the author(s). An editor may, however, take into account relationships of a manuscript immediately under consideration to others previously or concurrently offered by the same author(s).

2. An editor should consider manuscripts submitted for publication with all reasonable speed.

3. The sole responsibility for acceptance or rejection of a manuscript rests with the editor. Responsible and prudent exercise of this duty normally requires that the editor seek advice from reviewers, chosen for their expertise and good judgment, as to the quality and reliability of manuscripts submitted for publication. In reaching a final decision, the editor should also consider additional factors of editorial policy.

4. The editor and members of the editor's staff should not disclose any information about a manuscript under consideration to anyone other than those from whom professional advice is sought. (However, an editor who solicits, or otherwise arranges beforehand, the submission of manuscripts may need to disclose to a prospective author the fact that a relevant manuscript by another author has been received or is in preparation.) After manuscripts have been accepted for publication, the editor and members of the editor's staff may disclose or publish manuscript titles and authors' names, but no more than that unless the author's permission has been obtained.

5. An editor should respect the intellectual independence of authors.

6. Editorial responsibility and authority for any manuscript authored by an editor and submitted to the editor's journal should be delegated to some other qualified person, such as another editor of that journal or a member of its Editorial Advisory Board. Editorial consideration of the manuscript in any way or form by the author-editor would constitute a conflict of interest, and is therefore improper.

7. Unpublished information, arguments, or interpretations disclosed in a submitted manuscript should not be used in an editor's own research except with the consent of the author. However, if such information indicates that some of the editor's own research is unlikely to be profitable, the editor could ethically discontinue the work. When a manuscript is so closely related to the current or past research of an editor, as to create a conflict of interest, the editor should arrange for some other qualified person to take editorial responsibility for that manuscript. In some cases, it may be appropriate to tell an author about the editor's research and plans in that area.

8. If an editor is presented with convincing evidence that the main substance or conclusions of a report published in an editor's journal are erroneous, the editor should facilitate publication of an appropriate report pointing out the error, and if possible, correcting it. The report may be written by the person who discovered the error, or by an original author.

B. Ethical Obligations of Authors

1. An author's central obligation is to present an accurate account of the research performed as well as an objective discussion of its significance.

2. An author should recognize that journal space is a precious resource created at considerable cost. An author therefore has an obligation to use it wisely and economically.

3. A primary research report should contain sufficient detail and reference to public sources of information to permit the author's peers to repeat the work.

4. An author should cite those publications that have been influential in determining the nature of the reported work, and that will guide the reader quickly to the earlier work that is essential for understanding the present investigation. Except in a review, citation of work that will not be referred to in the reported research should be minimized.

5. Any unusual hazards inherent in the chemicals, equipment, or procedures used in an investigation should be clearly identified in a manuscript reporting the work.

6. Fragmentation of research reports should be avoided. A scientist who has done extensive work on a system or group of related systems should organize publication so that each report gives a well-rounded account of a particular aspect of the general study. Fragmentation consumes journal space excessively and unduly complicates literature searches. The convenience of readers is served if reports on related studies are published in the same journal, or in a small number of journals.

7. In submitting a manuscript for publication, an author should inform the editor of related manuscripts that the author has under editorial consideration or in press. The relationships of such manuscripts to the one submitted should be indicated.

8. It is in general inappropriate for an author to submit manuscripts describing essentially the same research to more than one journal of primary publication. However, there are exceptions as follows: (a) resubmission of a manuscript rejected by or withdrawn from publication in one journal; (b) submission of overlapping work to a second journal in another field, if workers in the other field are unlikely to see the article published in the first journal, providing that both editors are informed; and (c) submission of a manuscript for a full paper expanding on a previously published brief preliminary account (a "communication" or "letter") of the same work.

9. An author should identify the source of all information quoted or offered, except that which is common knowledge. Information obtained privately, as in conversation, correspondence, or discussion with third parties, should not be used or reported in the author's work without explicit permission from the investigator with whom the information originated. Information obtained in the course of confidential services, such as refereeing manuscripts or grant applications, should be treated similarly.

10. An experimental or theoretical study may sometimes justify criticism, even severe criticism, of the work of another scientist. When appropriate, such criticism may be offered in published papers. However, in no case is personal criticism considered to be appropriate.

11. The co-authors of a paper should be all those persons who have made significant scientific contributions to the work reported and who share responsibility and accountability for the results. Other contributions should be indicated in a footnote or an "Acknowledgments" section. An administrative relationship to the investigation does not of itself qualify a person for co-authorship (but occasionally it may be appropriate to acknowledge major administrative assistance). Deceased persons who meet the criterion for inclusion as co-authors should be so included, with a footnote reporting date of death. No fictitious name should be listed as an author or co-author. The author who submits a manuscript for publication accepts the responsibility of having included as co-authors all persons appropriate and none inappropriate. The submitting author should have sent each living co-author a draft copy of the manuscript and have obtained the co-author's assent to co-authorship of it.

C. Ethical Obligations of Reviewers of Manuscripts

1. Inasmuch as the reviewing of manuscripts is an essential step in the publication process, and therefore in the operation of the scientific method, every scientist has an obligation to do a fair share of reviewing.

2. A chosen reviewer who feels inadequately qualified to judge the research reported in a manuscript should return it promptly to the editor.

3. A reviewer (or referee) of a manuscript should judge objectively the quality of the manuscript, of its experimental and theoretical work, of its interpretations and its exposition, with due regard to the maintenance of high scientific and literary standards. A reviewer should respect the intellectual independence of the authors.

4. A reviewer should be sensitive to the appearance of a conflict of interest when the manuscript under review is closely related to the reviewer's work in progress or published. If in doubt, the reviewer should return the manuscript promptly without review, advising the editor of the conflict of interest or bias. Alternatively, the reviewer may wish to furnish a signed review stating the reviewer's interest in the work, with the understanding that it may, at the editor's discretion, be transmitted to the author.

5. A reviewer should not evaluate a manuscript authored or co-authored by a person with whom the reviewer has a personal or professional connection if the relationship would bias judgment of the manuscript.

6. A reviewer should treat a manuscript sent for review as a confidential document. It should neither be shown to nor discussed with others except, in special cases, persons from whom specific advice may be sought; in that event, the identities of those consulted should be disclosed to the editor.

7. Reviewers should explain and support their judgments adequately so that editors and authors may understand the basis of their comments. Any statement that an observation, derivation, or argument had been previously reported should be accompanied by the relevant citation. Unsupported assertions by reviewers (or by authors in rebuttal) are of little value and should be avoided.

8. A reviewer should be alert to failure of authors to cite relevant work by other scientists, bearing in mind that complaints that the reviewer's own research was insufficiently cited may seem self-serving. A reviewer should call to the editor's attention any substantial similarity between the manuscript under consideration and any published paper or any manuscript submitted concurrently to another journal.

9. A reviewer should act promptly, submitting a report in a timely manner. Should a reviewer receive a manuscript at a time when circumstances preclude prompt attention to it, the unreviewed manuscript should be returned immediately to the editor. Alternatively, the reviewer might notify the editor of probable delays, and propose a revised review date.

10. Reviewers should not use or disclose unpublished information, arguments, or interpretations contained in a manuscript under consideration, except with the consent of the author. If this information indicates that some of the reviewer's work is unlikely to be profitable, the reviewer, however, could ethically discontinue the work. In some cases, it may be appropriate for

the reviewer to write the author, with copy to the editor, about the reviewer's research and plans in that area.

D. Ethical Obligations of Scientists Publishing Outside the Scientific Literature

1. A scientist publishing in the popular literature has the same basic obligation to be accurate in reporting observations and unbiased in interpreting them as when publishing in a scientific journal.

2. Inasmuch as laymen may not understand scientific terminology, the scientist may find it necessary to use common words of lesser precision to increase public comprehension. In view of the importance of scientists' communicating with the general public, some loss of accuracy in that sense can be condoned. The scientist should, however, strive to keep public writing, remarks, and interviews as accurate as possible, consistent with effective communication.

3. A scientist should not proclaim a discovery to the public unless the experimental, statistical, or theoretical support for it is of strength sufficient to warrant publication in the scientific literature. An account of the experimental work and results that support a public pronouncement should be submitted as quickly as possible for publication in a scientific journal. Scientists should, however, be aware that extensive disclosure of research in the public press might be considered by a journal editor as equivalent to a preliminary communication in the scientific literature.

Appendix III

Element Symbols, Atomic Numbers, and Atomic Weights

Element	Symbol	Atomic Number	Atomic Weight
Actinium	Ac	89	227.03[a]
Aluminum	Al	13	26.98
Americium	Am	95	(243)
Antimony	Sb	51	121.75
Argon	Ar	18	39.95
Arsenic	As	33	74.92
Astatine	At	85	(210)
Barium	Ba	56	137.33
Berkelium	Bk	97	(247)
Beryllium	Be	4	9.01
Bismuth	Bi	83	208.98
Boron	B	5	10.81
Bromine	Br	35	79.90
Cadmium	Cd	48	112.41
Calcium	Ca	20	40.08
Californium	Cf	98	(251)
Carbon	C	6	12.01
Cerium	Ce	58	140.12
Cesium	Cs	55	132.90
Chlorine	Cl	17	35.45
Chromium	Cr	24	51.996
Cobalt	Co	27	58.93
Copper	Cu	29	63.55
Curium	Cm	96	(247)
Dysprosium	Dy	66	162.50
Einsteinium	Es	99	(252)
Element 106		106	(263)
Element 107		107	(262)

Element	Symbol	Atomic Number	Atomic Weight
Element 108		108	(265)
Element 109		109	(266)
Erbium	Er	68	167.26
Europium	Eu	63	151.96
Fermium	Fm	100	(257)
Fluorine	F	9	18.998
Francium	Fr	87	(223)
Gadolinium	Gd	64	157.25
Gallium	Ga	31	69.72
Germanium	Ge	32	72.59
Gold	Au	79	196.97
Hafnium	Hf	72	178.49
Hahnium	Ha	105	(262)
Helium	He	2	4.00
Holmium	Ho	67	164.93
Hydrogen	H	1	1.01
Indium	In	49	114.82
Iodine	I	53	126.90
Iridium	Ir	77	192.22
Iron	Fe	26	55.85
Krypton	Kr	36	83.80
Lanthanum	La	57	138.90
Lawrencium	Lr	103	(260)
Lead	Pb	82	207.2
Lithium	Li	3	6.94
Lutetium	Lu	71	174.97
Magnesium	Mg	12	24.30
Manganese	Mn	25	54.94
Mendelevium	Md	101	(258)
Mercury	Hg	80	200.59
Molybdenum	Mo	42	95.94
Neodymium	Nd	60	144.24
Neon	Ne	10	20.18
Neptunium	Np	93	237.05[a]
Nickel	Ni	28	58.69
Niobium	Nb	41	92.91
Nitrogen	N	7	14.01
Nobelium	No	102	(259)
Osmium	Os	76	190.2
Oxygen	O	8	15.999
Palladium	Pd	46	106.42
Phosphorus	P	15	30.97

Element	Symbol	Atomic Number	Atomic Weight
Platinum	Pt	78	195.08
Plutonium	Pu	94	(244)
Polonium	Po	84	(209)
Potassium	K	19	39.1
Praseodymium	Pr	59	140.91
Promethium	Pm	61	(145)
Protactinium	Pa	91	231.04[a]
Radium	Ra	88	226.02[a]
Radon	Rn	86	(222)
Rhenium	Re	75	186.21
Rhodium	Rh	45	102.90
Rubidium	Rb	37	85.47
Ruthenium	Ru	44	101.07
Rutherfordium	Rf	104	(261)
Samarium	Sm	62	150.36
Scandium	Sc	21	44.96
Selenium	Se	34	78.96
Silicon	Si	14	28.08
Silver	Ag	47	107.87
Sodium	Na	11	22.99
Strontium	Sr	38	87.62
Sulfur	S	16	32.06
Tantalum	Ta	73	180.95
Technetium	Tc	43	(98)
Tellurium	Te	52	127.60
Terbium	Tb	65	158.92
Thallium	Tl	81	204.38
Thorium	Th	90	232.04[a]
Thulium	Tm	69	168.93
Tin	Sn	50	118.69
Titanium	Ti	22	47.88
Tungsten	W	74	183.85
Uranium	U	92	238.03
Vanadium	V	23	50.94
Xenon	Xe	54	131.29
Ytterbium	Yb	70	173.04
Yttrium	Y	39	88.91
Zinc	Zn	30	65.38
Zirconium	Zr	40	91.22

NOTE: This table is based on 1979 IUPAC atomic weights of the elements.
A value given in parentheses denotes the mass number of the longest-lived isotope.
[a] Atomic weight of most commonly available long-lived isotope.

Appendix IV

Symbols for Commonly Used Physical Quantities

This list is based on IUPAC recommendations. Where more than one symbol is given for the same physical quantity, the alternatives may be chosen in cases of conflict.

Physical Quantity	Symbol	SI Unit
Space and Time		
Cartesian space coordinates	x, y, z	m
Position vector	$\mathbf{r}$	m
Length	l	m
Height	h	m
Thickness, distance	d, δ	m
Radius	r	m
Diameter	d	m
Path, length of arc	s	m
Area	A, S, A_S	m²
Volume	V	m³
Plane angle	$\alpha, \beta, \gamma, \theta, \phi$	1, rad
Solid angle	ω, Ω	1 sr
Time	t	s
Frequency	ν, f	Hz
Circular frequency $(=2\pi\nu)$	ω	s^{-1}, rad s^{-1}
Characteristic time interval, relaxation time, time constant	τ	s
Velocity	$\mathbf{v, u, w, c}$	m s^{-1}
Acceleration	$\mathbf{a}$	m s^{-2}
Mechanics		
Mass	m	kg
Reduced mass	μ	kg
(Mass) density	ρ	kg m^{-3}
Relative density	d	1

Physical Quantity	Symbol	SI Unit
Specific volume	v	$m^3\ kg^{-1}$
Moment of inertia	I	$kg\ m^2$
Momentum	$\mathbf{p}$	$kg\ m\ s^{-1}$
Angular momentum	$\mathbf{L}$	$kg\ m^2\ s^{-1}\ rad(= J\ s)$
Force	$\mathbf{F}$	$N\ (= kg\ m\ s^{-2})$
Weight	$\mathbf{G, W}$	N
Moment of force	M	$N\ m$
Pressure	p	Pa
Surface tension	γ, σ	$N\ m^{-1},\ J\ m^{-2}$
Energy	E	J
Potential energy	$E_p,\ V,\ \Phi$	J
Kinetic energy	$E_k,\ T,\ K$	J
Work	$w,\ W$	J
Hamiltonian function	H	J
Power	P	W

General Chemistry

Number of entities	N	1
Amount of substance	n	mol
Molar mass	M	$kg\ mol^{-1}$
Relative molar mass, molecular weight	M_r	1
Relative atomic mass, atomic weight	A_r	1
Molar volume	V_m	$m^3\ mol^{-1}$
Mass fraction	w	1
Volume fraction	ϕ	1
Mole fraction	x	1
Partial pressure of substance B	p_B	Pa
Number concentration, number density of entities	$C,\ n$	m^{-3}
Amount concentration, concentration	c	$mol\ m^{-3}$
Molality	$m,\ b$	$mol\ kg^{-1}$
Surface concentration	Γ	$mol\ m^{-2}$
Stoichiometric coefficient	ν	1
Extent of reaction	ξ	mol
Degree of dissociation	α	1

Chemical Kinetics

Rate of conversion	$\dot{\xi}$	$mol\ s^{-1}$
Rate of concentration change of substance B (through chemical reaction)	$r_B,\ v_B$	$mol\ m^{-3}\ s^{-1}$

Physical Quantity	Symbol	SI Unit
Rate of reaction	v	$mol\ m^{-3}\ s^{-1}$
Overall order of reaction	n	1
Rate constant, rate coefficient	k	$(m^3\ mol^{-1})^{n-1}\ s^{-1}$
Half-life	$t_{1/2}$	s
Preexponential factor	A	$(m^3\ mol^{-1})^{n-1}\ s^{-1}$
Energy of activation	E	$J\ mol^{-1}$
Collision (reaction) cross section	$\sigma_{(r)}$	m^2
Collision frequency (of a particle A)	Z_A	s^{-1}
(Total) collision number of particles A and B per time and volume	Z_{AB}	$m^{-3}\ s^{-1}$
Collision frequency factor	z_{AB}	$m^{-3}\ s^{-1}\ mol^{-1}$
Photochemical yield, quantum yield	ϕ	1

Atoms and Molecules

Nucleon number, mass number	A	1
Proton number, atomic number	Z	1
Neutron number	N	1
Electron mass	m_e	kg
Proton mass	m_p	kg
Atomic mass constant (unified atomic mass unit)	m_u	kg
Elementary charge (of a proton)	e	C
Ionization energy	E_i, I	J
Principal quantum number	n	1
Orbital quantum number	l, L	1
Orbital quantum number component	m_l, m_L	1
Electron spin quantum number	s, S	1
Electron spin quantum component	m_s, m_S	1
Total angular momentum quantum number	j, J	1
Total angular momentum component	m_j, m_J	1
Nuclear spin quantum number	I	1
Nuclear spin quantum component	M_I	1
Total term	T	m^{-1}
Electronic term	T_e	m^{-1}

Physical Quantity	Symbol	SI Unit
Vibrational term	G	m^{-1}
Rotational term	F	m^{-1}
Rotational constants	A, B, C	m^{-1}
Vibrational quantum number	v	1
Magnetogyric ratio	γ	$C\ kg^{-1}$
g-factor	g	1
g-factor	g	1
Larmor frequency	ω_L	s^{-1}
Quadrupole moment	Q, Θ, eQ	$C\ m^2$
Decay constant	λ	s^{-1}

Electricity and Magnetism

Electric charge	Q	C
Charge density	ρ	$C\ m^{-3}$
Electric current	I	A
Electric current density	j	$A\ m^{-2}$
Electric potential	V, ϕ	V
Electric potential difference, voltage	$U, \Delta V, \Delta\phi$	V
Electric field strength	$\mathbf{E}$	$V\ m^{-1}$
Electric displacement	$\mathbf{D}$	$C\ m^{-2}$
Capacitance	C	F
Permittivity	ϵ	$F\ m^{-1}$
Relative permittivity	ϵ_r	1
Dielectric polarization	$\mathbf{P}$	$C\ m^{-2}$
Electric susceptibility	χ_e	1
Polarization (of a particle)	α	$m^2\ C\ V^{-1}$
Electric dipole moment	$\mathbf{p}, \mathbf{p}_e$	$C\ m$
Magnetic flux density, magnetic induction	$\mathbf{B}$	T
Magnetic flux	Φ	Wb
Magnetic field strength, inductance	$\mathbf{H}$	$A\ m^{-1}, H$
Permeability	μ	$H\ m^{-1}$
Relative permeability	μ_r	1
Magnetization, magnetic moment	$\mathbf{M}$	$A\ m^{-1}$
Magnetic susceptibility	χ	1
Electromagnetic moment	$\mu, \mathbf{m}$	$A\ m^2, J\ T^{-1}$
Resistance	R	Ω
Conductance	G	S
Resistivity	ρ	$\Omega\ m$
Conductivity	κ	$S\ m^{-1}$
Self-inductance	L	H

Physical Quantity	Symbol	SI Unit
Thermodynamics and Statistics		
Heat	q, Q	J
Work	w, W	J
Thermodynamic temperature	T	K
Celsius temperature	t, θ	°C
Internal energy	U	J
Enthalpy	H	J
Entropy	S	$J\,K^{-1}$
Gibbs energy (function)	G	J
Helmholtz energy (function)	A	J
Heat capacity	C	$J\,K^{-1}$
Specific heat capacity	c	$J\,K^{-1}\,kg^{-1}$
Molar heat capacity	C_m	$J\,K^{-1}\,mol^{-1}$
Ratio C_p/C_v	γ	1
Joule–Thomson coefficient	μ	$K\,Pa^{-1}$
Isothermal compressibility	κ	Pa^{-1}
Pressure coefficient	β	$Pa\,K^{-1}$
Cubic expansion coefficient	α	K^{-1}
Chemical potential	μ	$J\,mol^{-1}$
Affinity of a reaction	$A, \mathscr{A}$	$J\,mol^{-1}$
Fugacity	$f, \widetilde{p}$	Pa
Osmotic pressure	Π	Pa
Activity (relative activity)	a	1
Activity coefficient, mole fraction basis	f	1
Activity coefficient, molality basis	γ	1
Activity coefficient, concentration basis	y	1
Osmotic coefficient	ϕ	1
Equilibrium constant, concentration basis	K_c	$(mol\,m^{-3})^{\Sigma v}$
Equilibrium constant, pressure basis	K_p	$Pa^{\Sigma v}$
Statistical weight	g	1
Partition function (particle)	q, z	1
Canonical ensemble partition function	Q, Z	1
Microcanonical ensemble partition function	Ω	1
Grand (canonical ensemble) partition function	Ξ	1
Symmetry number	σ, s	1
Characteristic temperature	Θ	K

Physical Quantity	Symbol	SI Unit
Radiation		
Wavelength	λ	m
Wavenumber	$\sigma, \tilde{v}$	m^{-1}
Radiant energy	Q, W	J
Radiant intensity	I	$W\ sr^{-1}$
Emissivity, emittance	ϵ	1
Absorptance	α	1
Transmittance	τ	1
Linear decadic absorption coefficient	a	m^{-1}
Molar decadic absorption coefficient	ϵ	$m^2\ mol^{-1}$
Refractive index	n	1
Molar refraction	R_m	$m^3\ mol^{-1}$
Angle of optical rotation	α	1, rad
Electrochemistry		
Charge number of an ion	z	1
Ion strength	I	$mol\ kg^{-1}$
Electromotive force	E	V
Electrochemical potential	$\tilde{\mu}$	$J\ mol^{-1}$
Electrode potential	E	V
pH	pH	1
Electrolytic conductivity	κ	$S\ m^{-1}$
Molar conductivity (of an electrolyte)	Λ	$S\ m^2\ mol^{-1}$
Molar conductivity (of an ion), ionic conductivity	λ	$S\ m^2\ mol^{-1}$
Electric mobility	u	$m^2\ V^{-1}\ s^{-1}$
Transport number	t	1
Transport Properties		
Flux of a quantity x	J_x, J	(varies)
Mass flow rate	$q_m, \dot{m}$	$kg\ s^{-1}$
Volume flow rate	$q_v, \dot{V}$	$m^3\ s^{-1}$
Heat flow rate	ϕ	W
Thermal conductivity	λ, k	$W\ m^{-1}\ K^{-1}$
Coefficient of heat transfer	h	$W\ m^{-2}\ K^{-1}$
Thermal diffusivity	a	$m^2\ s^{-1}$
Diffusion coefficient	D	$m^2\ s^{-1}$
Thermal diffusion coefficient	D_T	$m^2\ s^{-1}$
Viscosity	η, μ	Pa s
Kinematic viscosity	v	$m^2\ s^{-1}$

Appendix V

Hints to the Typist

This appendix does not apply to ACS Symposium Series manuscripts. Please refer to special instructions provided by the ACS Books Department.

Arranging the Manuscript

Number all pages of a manuscript consecutively, including the title page and lists of footnotes, references, tables, captions, and figures. If you insert or remove a page after the final copy has been prepared, number succeeding pages accordingly. References, tables, captions, and figures should be placed in that order, at the end of the paper. Do not put each caption on a single page; group them on one or more separate pages. Submit three good copies of the manuscript. A set of illustrations must accompany each copy: one set of originals, stats, or glossy prints and two sets of photocopies.

Typing the Manuscript

Type or print out manuscripts double-spaced, on one side only, on 8½- × 11-inch (22- × 28-cm) heavy-duty, white bond paper. Double space ALL copy: references, footnotes, tables, abstract, and figure captions as well as regular text. Leave a liberal margin, 1 inch (2 cm), on all four sides of each page. If you use a word processor or computer, use a letter quality (or character) printer or a comparably high-quality dot matrix printer. Many dot matrix printers produce output that is not dark enough, does not provide sufficient distinction between baseline and subscript characters, has poor intercharacter spacing, and lacks adequate character resolution. No printer having any of these limitations qualifies as a high-quality printer. You may also use a laser printer if the output is high-quality.

The final copy should be as nearly letter perfect as possible. If a correction must be made, cross out the error and type the correct version above it. You may also use correction tape or correction fluid. Overtyping an incorrect letter with a correct letter gives the printer no indication of which letter to print. Do not type in margins or below the lines, and do not attach slips of paper to the pages. Retype any page needing lengthy insertions. Extensive handwritten revisions are not acceptable.

Mathematical Patterns

Equations   The starting place in typing an equation, a fraction, or almost any pattern is called the "main line". The major symbols (=, +, -, <, >, etc.), punctuation, parentheses and brackets, and the equation number should be typed on this line. The required pattern is then typed either above or below this level.

Displayed equations (those that cannot fit on one line) should be separated from the text by a triple space below the lowest point. All symbols, Greek letters, signs, large parentheses, braces, and other special characters for which typewriter or print wheel elements are unavailable should be traced on the copy in black ink or pressed on with dry transfer type.

Handwritten directions identifying special characters should be made in the margin in pencil. The notation "use throughout" will aid both editor and printer.

Superscripts and Subscripts   Superscripts (also referred to as superiors or exponents) and subscripts (often called inferiors) may be single letters, Greek letters, numbers, mathematical symbols, or groups of these (multiple scripts) that form complex patterns.

Superscripts and subscripts belong specifically to a single character or to a pattern of many characters on the main line. Therefore, all superscripts and subscripts should be typed as closely as possible to the character or pattern to which they belong, but superscripts and subscripts that belong to the same character should not be typed directly above one another.

Single superscripts are typed slightly above the main line character and single subscripts slightly below the main line character. To line up all superscripts and subscripts, first type the main line, leaving space for the superscripts and subscripts, then go back and type them in. There may be superscripts and subscripts to superscripts, and superscripts and subscripts to subscripts. Be careful that these extra scripts do not wander too far from the pattern to which they belong.

$$K_{sp} = (\gamma_{+}m_{+})^{a}(\gamma_{-}m_{-})^{b} = m_{+}^{a}m_{-}^{b}(\gamma_{\pm})^{a+b}$$

$$\tilde{v}^{E} = (\tilde{v}^{0})^{7/3}[4/3 - (\tilde{v}^{0})^{1/3}]^{-1}[\tilde{T} - \tilde{T}^{0}]$$

In typing thermodynamic symbols, such as S°_0 and H°_0, use the degree sign as superscript and zero as the subscript. Remember that the scripts should not be one above the other. Never use the lower-case letter *o* to represent the cipher zero.

Fractions Wherever possible, fractions should be typed on one line with a slant (or solidus) instead of a horizontal fraction bar.

$$(2\underline{D}_z)^{1/2} \qquad\qquad \underline{xy}/\underline{z}$$

Complex fractions should be typed with a horizontal bar. The fraction bar is made with the underscore key and is exactly as long as the longer term. The smaller term is then centered on this fraction bar.

$$\log \left[\frac{(\underline{k}_{1\underline{c}}/\underline{k}_{1\underline{t}})_1}{(\underline{k}_{1\underline{c}}/\underline{k}_{1\underline{t}})_2} \right] = \frac{\Delta \underline{E}}{2.303\underline{R}} \left[\frac{1}{\underline{T}_2} - \frac{1}{\underline{T}_1} \right]$$

Notations The summation symbol Σ, the product symbol Π, and the integral symbol $\int$ may be drawn or dry transfer type may be used. The upper limits (above the symbol) and the lower limits (below the symbol) should be typed.

$$\underline{i}_{\underline{m}} = \underline{i} \sum_{\underline{K}=1}^{\underline{m}-1} \alpha_{\underline{k}\underline{m}} \underline{V}_{\underline{k}} \qquad\qquad \prod_{\underline{k}=0}^{\underline{m}-1} (1 - \underline{k}_{\underline{r}}^{\underline{m}})$$

Chemical Bonds

Use the "equals" sign for a double bond and a hyphen for a single bond. Never space around bond lines.

$$H_2C-CH=CH-CH_2$$

Chemical Reactions

The alignment and spacing of single-line chemical reactions are the same as for mathematical equations. The plus and arrow have one space before and after. When an equation must be carried over to a second line, break it at the arrow, and keep the arrow on the top line.

$$3Ca^{2+} \;+\; 2PO_4^{3-} \longrightarrow \; Ca_3(PO_4)_2$$

$$Ni(NH_2CH_2CH_2CO_2)_2 \;+\; NH_2^- \longrightarrow$$

$$[Ni(NHCH_2CH_2CO_2)(NH_2CH_2CH_2CO_2)]^- \;+\; NH_3$$

Typing Material for Microform Supplements

Microform supplements are available only with journals, not books. Material submitted as supplementary must be camera-ready copy. Economy of space is the guiding principle; therefore, single-spaced copy is preferred. All characters should be clean, legible, and leave a clear, sharp impression. The paper should be a good quality, clean, as new as possible, and white, and should be 8½ × 11 inches (22 × 28 cm). If the material is typewritten, a good ribbon and error-free copy are essential. The ink should not be smudged; the typing should not penetrate the paper and should not show any erasures or corrections. If computer printouts are used, they must be clear and the type must be dark, with no broken letters. Use a letter quality (or character) printer, not dot matrix.

Tables should be grouped together to conserve space. Figures should be drafted on a small scale and mounted, in multiple, with accompanying captions on the same paper as the text. Original ink drawings are preferred, and stats (matte prints) should be sent rather than glossies. Because of the reproduction process involved, glossy prints may result in filmed material of nonuniform quality.

Each paper containing microform material should have an explanatory paragraph at the end, giving a brief description of the nature and the amount of the material, as follows: "Supplementary Material Available: Spectra (^{1}H NMR) for compounds **3–10** (3 pages). Ordering information is given on any current masthead page." Any modification necessary will be introduced by the Journals Editorial Department.

Microforms are processed to conform to the highest industry standards; however, the quality of the finished microfiche or microfilm can be no better than the source document from which it is filmed. To produce high-quality, legible microforms, the following guidelines are important.

Characters should be large enough to be read comfortably in original form. A microfilmed document is reduced in filming. Then it is blown up to its original size, which is most frequently 8½ × 11 inches (22 × 28 cm). This is also the average microform reader screen size. Therefore, in determining if the characters and text are of readable size, the answer lies in the source document to be microfilmed. Is it readable in its original form?

Whenever possible, all characters and text in the information area should be upright and right-reading, that is, reading from left to right across the 8½-inch (22-cm) width of the paper. Oversized material or material that is displayed on the length of the page should be used only when it cannot be presented in the standard format.

The contrast and density of all material should be adequate and of sufficient uniformity so that all information on the source document can be reproduced with such fidelity that its use will not be impaired.

The copy to be microfilmed should be as clean and unmarked as possible. Staples will leave marks, and smudges will also reproduce. Use paper clips to keep the copy together and correction fluid to remove extraneous marks. Do not use adhesive or other shiny tape on the surface of the document. The tape causes glare when the copy is being microfilmed.

Appendix VI

Proofreaders' Marks

On galleys, mark in the margins and use these symbols. For Greek letters, mathematical symbols, or other special symbols, draw them in the margin but also spell out the word and circle it.

Operational Signs

ℓ	Delete
◡	Close up; delete space
ℓ̲	Delete and close up
#	Insert space
¶	Begin new paragraph
no ¶	Run paragraphs together
□	One em space
]	Move right
[	Move left
] [	Center
⊓	Move up
⊔	Move down
=	Align horizontally
‖	Align vertically
(tr)	Transpose
(sp)	Spell out
(stet)	Let it stand
(fl ℓ)	Flush left
(fl rt)	Flush right
(ctr)	Center

Typographical Signs

(lc)	Lower case a capital letter
(cap)	Capitalize a lower-case letter
(sc)	Set in small capitals
(ital)	Set in italic type
(rom)	Set in roman type
(bf)	Set in boldface type
(wf)	Wrong font; set in correct type
∨	Superscript
∧	Subscript

Punctuation Marks

⌃	Insert comma
⌄	Insert apostrophe (or single quotation mark)
⌄⌄	Insert quotation marks
⊙	Insert period
?	Insert question mark
;	Insert semicolon
:	Insert colon
ꞁ=ꞁ	Insert hyphen
⊥/M	Insert em dash
⊥/N	Insert en dash

0917–0/86/0239$06.00/0

The photochemistry of α,β-unsaturated ketones has attracted much attention and is still a field field of current interest. Numerous examples of such photochemical transformations are well-documented for cyclic enones and dienones, including both cycloaddition reactions and rearrangements. For example, cyclopentenones *1* and *2* readily rearrange to cyclopropyl ketenes upon irradiation. Recently, the related cyclohexadienone/butadienyl ketene rearrangement has been shown to be a highly useful tool in the synthesis of natural products and macrocyclic lactones.

Whereas cis/trans isomerization, photodimerization, and [2 + 2] cycloadditions of acyclic α,β-unsaturated ketones are well-investigated photochemical transformations, comparatively little is documented concerning the photochemistry of such enones involving photodissociation, rearrangement or both. Clearly, the absence of ring strain lowers the reactivity toward bond cleavage and renders an initial Norrish type I reaction unlikely. Introduction of radical stabilizing groups in the α-position of the enone may, however, be expected to change the reactivity of the enone in favor of the photochemical α-cleavage and subsequent reactions derived from the resulting radical pairs.

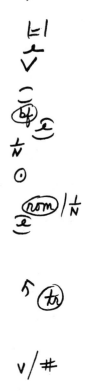

On manuscripts, you can mark in place, not in the margins. Some of the symbols differ, too. For example, to delete, simply strike through the word or words to be deleted. To capitalize a lower-case letter, use three scores below the letter. To lower case a capital letter, mark a slash through the letter. To transpose two words or letters, use a sideways **S**-shaped line that starts above one letter or word and continues below the one to be transposed. For small capital letters, use two scores below the letter. For boldface, use a wavy line below the letter or words. For superscripts and subscripts, draw the carat over or under the character itself. When you use "stet" on a manuscript, you must also put dots or short dashes under the copy that you wish to retain. When you use these marks, you need not explain them in the margin of the manuscript. They are standard.

The photochemistry of α,β-unsaturated ketones has attracted much attention and is still a field ~~field~~ of current interest. ¶ Numerous examples of such photochemical transformations are well-documented for cyclic enones and dienones, including both cycloaddition reactions and rearrangements. For example, cyclopentenones 1 and 2 readily rearrange to cyclopropyl ketenes upon irradiation. Recently, the related cyclohexadienone-butadienyl ketene rearrangement has been shown to be a highly useful tool in the synthesis of natural products and macrocyclic lactones. ²

Whereas *cis-trans* isomerization, photodimerization, and [2 + 2] cycloadditions of acyclic α,β-unsaturated ketones are well-investigated photochemical transformations, comparatively little is documented concerning the photochemistry of such enones involving photodissociation, rearrangement, or both. Clearly, the absence of ring strain lowers

the reactivity toward bond cleavage and renders
an initial Norrish type I reaction unlikely.
Introduction of radical stabilizing groups in
the α-position of the enone may, however, be
expected to change the reactivity of the enone
in favor of the photochemical α-cleavage and
subsequent reactions derived from the resulting radical pairs.

Bibliography

Chemical Nomenclature

IUPAC *Nomenclature of Inorganic Chemistry,* 2nd ed.; Butterworths: London, 1971. (U.S. distributor: Crane, Russak, New York.)

IUPAC *Nomenclature of Organic Chemistry, Sections A, B, C, D, E, F, and H;* Pergamon: Elmsford, NY, 1979.

The Merck Index: An Encyclopedia of Chemicals, Drugs, and Biologicals, tenth ed.; Merck: Rahway, NJ, 1983.

Nomenclature of Organic Compounds; Fletcher, John H.; Dermer, Otis C.; Fox, Robert B., Eds.; American Chemical Society: Washington, DC, 1974.

Banks, James E. *Naming Organic Compounds,* 2nd ed.; Saunders: Philadelphia, 1976.

USAN and the USP Dictionary of Drug Names; Griffiths, M. C.; Fleeger, C. A.; Miller, L. C., Eds.; United States Pharmacopeial Convention: Rockville, MD, 1984.

Ring System Handbook; Chemical Abstracts Service: Columbus, OH, 1984.

Biochemical Nomenclature and Related Documents; The Biochemical Society: London, 1978.

Weast, R. C. *Handbook of Chemistry and Physics,* 66th ed.; CRC: Cleveland, OH, 1985.

IUPAC "Manual of Symbols and Terminology for Physicochemical Quantities and Units"; *Pure Appl. Chem.* **1979,** *51,* 1-41.

Enzyme Nomenclature 1984: Recommendations of the Nomenclature Committee of the International Union of Biochemistry on the Nomenclature and Classification of Enzyme-Catalyzed Reactions; International Union of Biochemistry, 1984.

The SI

"Metric Editorial Guide", 4th ed.; American National Metric Council: Washington, DC, 1984.

"American National Standard Metric Practice"; ANSI Z210.1-1976; American National Standards Institute: New York, 1976.

Antoine, Valerie *Guidance for Using the Metric System. SI Version*; Society for Technical Communication: Washington, DC, 1976.

"The International System of Units (SI)"; Special Publication No. 330; National Bureau of Standards: Washington, DC, 1981.

Standard for Metric Practice; ASTM E 380-77; American Society for Testing and Materials: Philadelphia, 1977.

"Units of Measurement", ISO Standards Handbook 2, 2nd ed.; American National Standards Institute: New York, 1982.

Style Manuals

The Chicago Manual of Style, 13th ed.; The University of Chicago: Chicago, 1982.

U.S. Government Printing Office Style Manual; Government Printing Office: Washington, DC, 1984.

CBE Style Manual, 5th ed.; Council of Biology Editors: Bethesda, MD, 1983.

Handbook for AOAC Members, 5th ed.; Association of Official Analytical Chemists: Washington, DC, 1982.

Handbook and Style Manual; American Society of Agronomy, Crop Science Society of America, and Soil Science Society of America: Madison, WI, 1976.

Publication Manual of the American Psychological Association, 3rd ed.; American Psychological Association: Washington, DC, 1983.

Style Manual, 3rd ed.; American Institute of Physics: New York, 1978.

Swanson, Ellen *Mathematics into Type*; American Mathematical Society: Providence, RI, 1979.

Barclay, W. R.; Southgate, M. T.; Mays, R. W. for the American Medical Association *Manual for Authors and Editors*; Lange Medical: Los Altos, CA, 1981.

The McGraw-Hill Style Manual; Longyear, M., Ed.; McGraw-Hill: New York, 1983.

Guides for Writers and Editors

Skillin, M. E.; Gay, R. M. *Words into Type,* 3rd ed.; Prentice-Hall: Englewood Cliffs, NJ, 1974.

Strunk, W. S., Jr.; White, E. B. *The Elements of Style,* 3rd ed.; Macmillan: New York, 1979.

Berry, T. E. *The Most Common Mistakes in English Usage;* McGraw-Hill: New York, 1971.

Bernstein, T. M. *The Careful Writer: A Modern Guide to English Usage;* Atheneum: New York, 1965.

Williams, J. M. *Style: Ten Lessons in Clarity and Grace;* Scott, Foresman: Glenview, IL, 1981.

Judd, K. *Copyediting: A Practical Guide;* William Kaufmann: Los Altos, CA, 1982.

King, L. S. *Why Not Say It Clearly: A Guide to Scientific Writing;* Little, Brown: Boston, 1978.

Plotnik, A. *The Elements of Editing;* McMillan: New York, 1982.

van Leunen, M.-C. *A Handbook for Scholars;* Alfred A. Knopf: New York, 1979.

Ross-Larson, B. *Edit Yourself: A Manual for Everyone Who Works with Words;* N. W. Norton: New York, 1982.

Day, R. A. *How to Write and Publish a Scientific Paper;* ISI Press: Philadelphia, 1979.

O'Connor, M.; Woodford, F. P. *Writing Scientific Papers in English: An Else-Ciba Foundation Guide for Authors;* Elsevier: New York, 1975.

Perrin, P. G. *Writer's Guide and Index to English,* 6th ed.; Scott, Foresman: Glenview, IL, 1978.

Fowler, H. W. *A Dictionary of Modern English Usage,* 2nd ed.; Oxford University: New York, 1983.

Hodges, J. C.; Whitten, M. E. *Harbrace College Handbook,* 9th ed.; Harcourt, Brace, Jovanovich: New York, 1982.

Baker, S. *The Complete Stylist,* 2nd ed.; Thomas Y. Crowell: New York, 1972.

Rathbone, R. R. *Communicating Technical Information: A Guide to Current Uses and Abuses in Scientific and Engineering Writing;* Addison-Wesley: Reading, MA, 1972.

Copperud, Roy H. *American Usage and Style: The Consensus;* Van Nostrand-Reinhold: New York, 1980.

Flesch, R. *The ABC of Style: A Guide to Plain English;* Harper & Row: New York, 1966.

Style in Special Fields Other Than Chemistry

Jeffrey, C. *Biological Nomenclature,* 2nd ed.; Crane, Russak: New York, 1977.

Bergey's Manual of Determinative Bacteriology, 8th ed.; Buchanan, R. E.; Gibbons, N. E., Eds.; Williams & Wilkins: Baltimore, 1974.

International Code of Nomenclature of Bacteria. Bacteriological Code. Lapage, S. P.; Sneath, P. H. A.; Lessel, E. F.; Skerman, V. B. D.; Seeliger, H. P. R.; Clark, W. A., Eds.; American Society for Microbiology: Washington, DC, 1976.

Committee on Names of Fishes *A List of Common and Scientific Names of Fishes from the United States and Canada,* 4th ed.; American Fisheries Society: Bethesda, MD, 1980.

Committee on Common Names of Insects *Common Names of Insects and Related Organisms Approved by the Entomological Society of America;* Entomological Society of America; College Park, MD, 1982.

Handbook of Biochemistry and Molecular Biology, 3rd ed.; Fasman, G. D., Ed.; CRC: Cleveland, OH, 1976; Vol. 2.

Index

A

C

D

T

X

Production Editor: Anne T. Riesberg
Indexers: Meg Marshall and Karen McCeney
Proofreaders: Hilary Kanter and Deborah H. Steiner

Text and cover design and cover art by Pamela Lewis

Typesetting by Hot Type, Ltd., Washington, DC
Printing and binding by R. R. Donnelley & Sons Company, Harrisonburg, VA